"A great many factors, beginning with changes in the pattern of our weather, have been alerting us to the urgent need for a wholly newer approach to the problems of ecology. From a Christian point of view this is a situation which calls out for a new and creative affirmation of a truly Cosmic Christology. In this context the pioneering works of Teilhard de Chardin coming to us from the first half of the twentieth century, reveal a startling relevance, a still-unrecognised potential. Siôn Cowell's *Lexicon*, the result of a long and careful reading of the Teilhardian corpus, will surely prove of great importance in promoting a new discovery and a new appraisal of Teilhard's thought. In particular, the writer's close familiarity with the original French of Teilhard's writings, makes the book an invaluable tool for study and research, elucidating in English the complexities of language which Teilhard forged to express the newness of a vision, at once scientific and theological." *Canon A. M. (Donald) Allchin, Honorary Professor in the Department of Theology, University of Wales Bangor, Warden of the Community of the Sisters of Ty Mawr*

"This English-Language Lexicon of Teilhard de Chardin's writings will be of indispensable value in all future study of Teilhard in the English language. A new phase in Teilhard studies should soon make its appearance as we understand more clearly the insight into the evolutionary process that he has given us. This new resource will help immensely." *Thomas Berry, Past President, American Teilhard Association*

"This Lexicon is an essential resource for all serious students of Teilhard de Chardin, especially those who only have access to English translations of his works. The author has given us a comprehensive survey of all terms likely to be encountered, along with clear and lucid summaries of the meanings of these terms. In addition, he includes some background information on particular writers, theologians and mystics who have influenced Teilhard de Chardin, hence enabling the serious reader to gain a further window into the way Teilhard's thinking developed. This book will be valuable not just to specialists in Teilhard, but all students who are encountering his work for the first time." *Professor Celia E. Deane-Drummond, University College Chester*

"Readers of Teilhard will find Siôn Cowell's excellent lexicon very helpful and useful, not only for the seven hundred entries that clarify Teilhard's quite special vocabulary, terms and concepts, but also for the comprehensive bibliography and the concise and informative introduction to Teilhard's life and works. Certainly, this book belongs in every college and university library." *Robert Faricy SJ, Emeritus Professor of Spiritual Theology, Pontifical Gregorian University, Rome*

"Siôn Cowell's work on *The Teilhard Lexicon* has my full endorsement. This overview of the terminology, thought, and creative connections made by Teilhard is a remarkable contribution for those interested in understanding Pierre Teilhard de Chardin's work. *The Teilhard Lexicon* does not simply define and explicate specific terms in Teilhard's work, rather, Siôn Cowell has drawn from Teilhard's opus specific quotations which reference the definitions given in his work. In effect *The Teilhard Lexicon* can be used as both a study guide and a meditational handbook in Teilhard's vision of a universe in process of convergence towards a cosmic centre of unity." *Professor John A. Grim, President, American Teilhard Association, Department of Religion, Bucknell University*

"Readers of Teilhard have often been confused by his new words and new uses of old words. He had a highly original vision and sometimes had to stretch words to convey his meaning. To bring the vision to more readers Siôn Cowell spent years in developing *The Teilhard Lexicon.* Now those hesitant to begin reading Teilhard will have a new access to his thought and scholars will be able to appreciate the precision of his language. It will be a standard reference for years to come." *Professor Thomas M. King SJ, Georgetown University, Washington, D.C. Author of* Teilhard's Mysticism of Knowing *and* Teilhard de Chardin, *and co-editor of* The Letters of Teilhard de Chardin and Lucile Swan

"*The Teilhard Lexicon* is an excellent reference work for anyone who wants to understand the immense range, depth and complexity of Teilhard's thought. Here are listed the major terms of his specialised vocabulary, accompanied by definitions drawn from Teilhard's own works. Set in a wider explanatory context, the entries make thought-provoking reading and inviter further study of Teilhard's important oeuvre. Such a lexicon has not been available in the English-speaking world before, and it is thus a splendid addition to the existing secondary sources which facilitate the understanding of this seminal thinker of the twentieth century." *Professor Ursula King, Department of Theology, Bristol University*

The Teilhard Lexicon

Do Phronnsias Ní Ceallaigh IBVM

Anamchara dhilis*

* "A true soul friend."

The Teilhard Lexicon

Understanding the language, terminology and vision of the writings of Pierre Teilhard de Chardin

The First English-Language Dictionary of his writings

SIÔN COWELL

Brighton • Portland

2 4 6 8 10 9 7 5 3 1

First published in 2001 in Great Britain by
SUSSEX ACADEMIC PRESS
PO Box 2950
Brighton BN2 5SP

and in the United States of America by
SUSSEX ACADEMIC PRESS
5824 N.E. Hassalo St.
Portland, Oregon 97213-3644

Acknowledgments for permission to reproduce copyright material are detailed on pp. 231–2

British Library Cataloguing in Publication Data
A CIP catalogue record for this book is available from the British Library.

Library of Congress Cataloging-in-Publication Data
Cowell, Siôn.
The Teilhard lexicon : understanding the language, terminology, and vision of the writings of Pierre Teilhard de Chardin / Siôn Cowell.
p. cm.
Includes bibliographical references.
ISBN 1-902210-37-9 (alk. paper)
1. Teilhard de Chardin, Pierre—Dictionaries. I. Title.
B2430.T373 Z83 2001
194—dc21
2001017016

Typeset and designed by G&G Editorial, Brighton
Printed by T.J. International, Padstow, Cornwall
This book is printed on acid-free paper

Contents

Preface

Born on Sunday 1 May 1881 in Sarcenat in the French Auvergne, Pierre Teilhard de Chardin died on Easter Sunday, 10 April 1955, in New York. Much of his life was spent in virtual exile – often in remote corners of the world where his religious superiors thought he might cause the least theological disturbance.

Teilhard was deeply influenced by his background. From his father, he learned love of the Earth, from his mother, a great grand-niece of Voltaire (1694–1778), love of God. Both loves were to influence deeply his scientific work and his religious and mystical sense. But love of the earth and love of God appeared to contradict one another. How could he resolve the apparent conflict between the two loves? This was a question Teilhard was to address in more than fifty years of active apostolate around the world.

In 1899 he entered the Society of Jesus. He was to remain committed to the Society for the rest of his life. For Teilhard de Chardin, man of science and man of prayer, the Society of Jesus was an "order of pioneers", at the head of the vanguard (Gérard-Henry Baudry, *Ce que croyait Teilhard*, 1971, 114–15). His friend and biographer Bruno de Solages (1895–1983) sees him as "the greatest christian apologist since Pascal" (Bruno de Solages OCarm, *Teilhard de Chardin*, 1967, 390). For his confrère and last superior André Ravier (1905–99), "Teilhard is, above all, a religious, a son of St. Ignatius, a priest and a missionary" (André Ravier SJ, "Teilhard de Chardin et l'expérience mystique d'après ses notes intimes," *Cahiers Pierre Teilhard de Chardin*, N° 8, 1974, 212). He belongs to a generation of great Jesuits that includes Henri de Lubac (1896-1991), Yves de Montcheuil (1900-44), Auguste Valensin (1879-1953), Hans Urs von Balthasar (1905-88) and Victor Fontoynant (1880–1958).

Frequently misunderstood by friend and foe alike, in his lifetime his reputation was grounded in his palaeontological research. His name will always be linked with the discovery of Peking Man (*Homo erectus pekinensis*) in 1929. And his work was to earn him a well-deserved place in the world of science. In 1950 he was elected to the French Academy of Sciences. In 1965 and again in 1981 he was honored by UNESCO as a scientist, philosopher and religious thinker of world repute. He contributed much to the Church renewal and inter-faith dialogue that were to flow from the Second Vatican Council (1962–5). On the centenary of his birth in 1981 his status as a religious thinker was recognized at an international colloquium held at the Catholic Institute in Paris. Letters from Cardinal Agostino Casaroli (1914–98) on behalf of Pope John Paul II and from Jesuit General Pedro Arrupe (1907–91) confirmed his standing in the contemporary Church.

Evolution is the key to Teilhard's commitment as priest and scientist. As a student and young paleontologist before World War I (1908–12) he had become convinced of the truth of evolution. Evolution was a fact. Things had evolved. They had not been created as such. The evidence lay there in the earth, in the rocks, in the fossil record. John Henry Newman (1801–90), whose name must also be linked with Vatican II, occupies an important place in the development of Teilhard's thought. His *Essay on the Development of Christian Doctrine* (1845) influenced, not only Teilhard's thinking on the development of christian dogma, but also, by transposition, his views on cosmic evolution. Teilhard sees the evolutionary process having meaning and direction towards its ultimate consummation in what he calls "a pole of cosmic convergence" that he identifies with the Risen Christ – the Christ Pantocrator. For Teilhard the human species is presently the leading shoot of evolution. We alone know and know that we know. We alone can observe and reflect upon ourselves and upon the universe of which we are a product. We alone are capable of giving meaning to the universe as we know it. We are no longer simple spectators of evolution. We are increasingly capable of influencing for good or ill the future course of planetary and cosmic evolution. We have become co-creators with the creator.

Teilhard's biographer Claude Cuénot (1911–92) rightly calls him "a cosmo-mystic". But he was a mystic who happened to be a scientist who had, in addition, a remarkable gift of poetic imagery. He was steeped in the mystics. They spoke his language. They anticipated his fears. They responded to his hopes. He partakes of that love of creation that is characteristic of mystics down the ages. His spirituality draws heavily on the eastern or orthodox tradition as well as the western or catholic tradition to which he belongs. Both traditions deny the sharp distinction between the sacred and the secular that informed much of post-renaissance thinking in western Europe. The traditional word for mystical experience – contemplation (Lat. *contemplatio*) – means "seeing" or "gazing intently at." Hence emphasis on the visual aspect in all mystical writers. Ignatius of Loyola (1491–1556) speaks of "contemplation in action". He teaches in *The Spiritual Exercises* a way of seeing God in all things because God dwells in all things, "gives them being, preserves them, grants them growth, sensation, etc." Ignatius himself found God at the heart of all reality – in nature, in the city and in loving service to those in need. So, too, did Teilhard. "It was in the Far East, on the road already trodden by Francis Xavier, de Nobili and Ricci, that Teilhard de Chardin, man of science and man of prayer, was to realize the Ignatian ideal of contemplation in action" (Alain Guillermou, *St. Ignace et la Compagnie de Jésus*, 1958, 177).

Teilhard seeks, above all, to construct a synthesis or coherent whole bringing together reason and authority, matter and spirit and science and faith. Science and religion often appear in conflict. But this conflict is more apparent than real. They deal with different aspects of one and the same reality. Teilhard opts for coherence, not concordance, where concordance means trying to make science and religion fit together at all times and at all levels while coherence means meeting together at "a pole of common vision." For Teilhard, scientific language explains reality but religious language communicates it. He uses both. He speaks of his system as a "hyper-physics" – an attempt at a correlation between the views of science, philosophy and myth on origins and goals and the insights of theology on cosmic history. He is driven at all times by a passionate belief in the presence of God at the heart of creation – or, as he would prefer to say, evolutionary creation. He is at one with the Church Fathers and the medieval scholastics in speaking of a "*creatio continua*" – creation in process.

Teilhard cannot be locked into fixed categories. His approach is essentially transdisciplinary. Teilhard the scientist easily becomes Teilhard the christian apologist. Many of those who are unhappy with Teilhard do not like the way he easily crosses the boundaries of science, philosophy and theology into the less well-defined fields of poetry and mysticism. But this, in a very real sense, is the essence of Teilhard. This criticism is largely the consequence of his advocacy of wholism and his rejection of reductionism – theory that complex phenomena can only be understood in terms of something simpler. He believes "the whole" is always greater than "the sum of its parts". He sees his scientific knowledge illuminated by his mystical insights. The universe is a "divine milieu". He shares with Angela of Foligno (1401–64) the vision of a universe "filled with God". His scientific knowledge is used to express these mystical insights within an evolutionary framework. The words "enfolding" and "unfolding" are vital to understanding his thought. He sees the process of "enfolding" (Fr. *enroulement*) as more fundamental to the evolutionary process than the traditional "unfolding" (Fr. *déroulement*) of nineteenth century thinkers.

A physical "unfolding" of the universe is meaningless without what he sees as a psychic (or spiritual) "enfolding". Teilhard believes many people "are reluctant to recognize that evolution has a precise orientation." This reluctance is still apparent today. He seeks constantly to combat this reluctance. Evolution is a system of interrelated events, not a collection of independent and unrelated happenings. It is, by definition, wholistic.

Teilhard found himself driven to develop a language very much his own to express the breadth and richness of his thought. He readily recognized that his language is not easy. He had always intended to prepare his own lexicon. Early in 1951 he had jotted down a list of words including some of the most important teilhardian "keywords" many of which have since been adopted by scientists, philosophers, theologians and others. He was never able to complete his lexicon. This is unfortunate – not least because he frequently uses words in his own particular way. He reworks existing words. And he develops new words where none previously existed. His language reflects his culture and his vocation. His background is essentially scholastic. Claude Cuénot (1911–92) sees him as the last of the great scholastics. But if scholasticism is the starting point of teilhardian thought it is only the starting point. Other philosophical influences are evident, e.g. Leibnitz (1646–1716), Blondel (1861–1949). Anglo-Saxon influences are apparent from his early years in Hastings (1908-12) and, even more, from 1926 onwards when he was in almost daily contact with the international community in China where English was the lingua franca. Indeed, English soon became his second language. His vocabulary is enriched by anglicisms – despite the best efforts of friends like Bruno de Solages (1895–1983) to find French expressions to represent "keywords" such as "Golden Glow", "pattern", "pull", "push", "stage", "trend", "upward", etc. Teilhard uses language to express synthesis, to build bridges between disciplines. He seeks to develop a language that is intelligent but intelligible. His vocabulary is filled with Latin and Greek words and derivatives. But his approach is that of the wholistic scientist and religious thinker, not the Latinist or the Hellenist. Julian Huxley (1887-1975) tells us the linking of evolutionary biology with christian theology is Teilhard's "unique contribution to thought, enabling thousands of christians to accept the greatest discovery since Newton – Darwin's discovery of the evolutionary process as a fact and as a scientifically explicable phenomenon – and so pave the way for the eventual reconciliation of science and religion . . . " (Huxley, introduction, in Barbour, 1965, 8).

Teilhard has not always been well-served by his translators. This is not altogether their fault. The Ignatian undertones that resonate throughout his writings are not readily identifiable by those unacquainted with the language of St. Ignatius. But it has led to inaccuracics and omissions that frequently distort the authors intentions. When what Huxley calls

Teilhard's seminal work, *Le Phénomène humain* (1938–40), was first translated into English it was wrongly called "The Phenomenon of Man." This is wholly inconsistent with Teilhard's thinking that focuses, above all, on "the phenomenon". He speaks rather of "the human phenomenon" and "nothing but the phenomenon; but also, the whole of the phenomenon" (*The Human Phenomenon*, 1938–40, I, 1 E; 21 F): elsewhere he speaks of "the biological phenomenon", "the christian phenomenon", "the mystical phenomenon", "the religious phenomenon" or "the spiritual phenomenon". The translation published by Sussex Academic Press (1999) uses the correct English title: *The Human Phenomenon*. Other English translations have been revised to bring them into line with what Teilhard's original language and meaning. Users should consult the Select Bibliography for details of English translations.

The Teilhard Lexicon is inspired by the *Nouveau lexique Teilhard de Chardin* by Claude Cuénot (1968). It also draws on the *Nouvel index analytique* by Paul L'Archevêque (1972), the *Vocabulaire Teilhard* by Hubert Cuypers (1963) and the *Teilhard de Chardin-Lexikon* by Adolf Haas SJ (1971). But it is far from being a simple translation. It contains many new entries such as "apologetics", "catholicism", "differentiation", "ethics", "futurism", "morality", "personalism", "universalism", etc. There is not, as far as we know, any comparable lexicon in the English language. Furthermore, most of Teilhard's writings such as his collected works that have been translated into English are no longer in print. Other writings such as his important correspondence with Christophe Gaudefroy (1878-1971) or fellow-Jesuits like Auguste Valensin (1879–1953) and Henri de Lubac (1896–1991) have never been translated. *The Teilhard Lexicon* makes full use of this correspondence. It is hoped it will provide users of all ages with a useful tool for a fuller understanding of the richness and diversity of teilhardian thought.

A very special word of thanks for his help, advice and encouragement is due to my publisher, Anthony Grahame of Sussex Academic Press, without whom publication would have been impossible. It would have been impossible to prepare this Lexicon without the support of friends at home and abroad and the assistance and tolerance of my wife Caroline. Eugene McCarthy CP, superior at St. Joseph's Church, Avenue Hoche, Paris, in the mid-seventies, was instrumental in encouraging me to begin my teilhardian studies. I owe a particular debt of gratitude to Teilhard's friend and confrère, his companion during his enforced residence in Peiping during World War II, Pierre Leroy (1900–92), with whom I spent many happy and instructive hours at the Jesuit house in Versailles.

Above all, it is Teilhard himself who has become my inseparable companion on my own spiritual voyage of discovery.

How to use the Lexicon

The system of referencing employed in *The Teilhard Lexicon* allows the user to trace sources without constantly having to refer back to the *Select Bibliography*. For this purpose, full references to the book, essay, letter or other source from which the quotation is taken are given at the end of each quotation followed by the year in which they were written. Roman numerals indicate the volume number used in the French edition of Teilhard's collected works (Seuil, 1955–76); full details of these and their English translations are found in the Select Bibliography. English page references are indicated by an "E" and French by an "F".

Example: "Above" in *The Teilhard Lexicon* under the reference "*The Religious Value of Research*, 1947, IX, 203 E; 261 F." The reference is taken from the essay called *The Religious Value of Research* written in 1947 and found in Volume IX (called *Science and Christ* in English) of Teilhard's collected works on page 203 in the English translation and page 261 in the French.

Siôn Cowell
Beaumaris
Feast of Saint Maelog
31 December 2000

About the Editor

Formerly deputy secretary general of an international non-governmental organization in Paris, 1975–89, Siôn Cowell has travelled and lectured widely throughout Western and Central Europe. He has worked on assignments with, inter alia, the *International Juridical Organization* (IJO), *the Organization for Economic Cooperation and Development* (OECD), the *Council of Europe*, the *United Nations Conference on Trade and Development* (UNCTAD), the *Stockholm Environment Institute* and the *Geneva Association.*

In Paris he was privileged to meet many of those who had known and worked with Pierre Teilhard de Chardin including his secretary Jeanne-Marie Mortier (1892–1982) and his friend and confrère Pierre Leroy (1900–92). He is now an independent researcher, writer and lecturer specializing in the evolutionary thought of Teilhard de Chardin.

President of the *European Teilhard de Chardin Centre*, 1991–93 and 1998–2000, he has been Chairman of *The British Teilhard Association* since 1991. He has long felt the need for a lexicon for English-speaking students and researchers. The present text is the product of more than ten years' research.

The Teilhard Lexicon

ABOVE *see also* **God-above**

Upwards movement of human beings towards the above. Spiritual ascent of humanity towards Christ in whom it will find its completion and its consummation when "God will be all in all" (1 Cor. 15:28).

> There can be no really living christian faith if it does not reach and lift up, in its ascending movement, the totality of human spiritual dynamism (the total of the "*anima naturaliter christiana*"). And, furthermore, no faith in the human is psychologically possible if the evolutionary future of the world does not meet up, in the transcendent, with some focal point of irreversible personalization. In short, it is impossible to rise above without moving ahead – or to progress ahead without rising towards the above. (*The Religious Value of Research*, 1947, IX, 203 E; 261 F)

ABSOLUTE *see also* **Transphenomenal**

> The need to possess "something absolute" in everything was the axis of my interior life from childhood . . . From then on I had the irresistible (and vitalizing yet relaxing) need to find lasting rest in something tangible and definitive. (*My Universe*, 1918, XIII, 197–8 E; XII, 296 F)

ABSTRACTION, UNITY OF *see also* **West, Road of the**

Mystical result obtained by the Road of the West.

> For eastern mysticism, the resolution of the multiple into the one is brought about by suppressing the multiple, unity having nothing in common with the multiple from which it must be separated (*maya*) . . . This is the difference between unity of abstraction and unity of richness (of concentration) where the elements find unity through perfection of all they possess. (Lecture, 18 January 1933, Claude Cuénot, *Teilhard de Chardin*, 140–1 E; 174–5 F)

ACT *see also* **Action**

> (a) To act worthily and usefully . . . is to be united. But to be united is to be transformed into something greater than the self. Thus to act is to finally leave the material, the immediate, the self-centered, to advance into the universal reality that is in process of being born. (*My Universe*, 1924, IX, 69 E; 97–8 F)
>
> (b) To act (that is, to apply our will to achieving some progress). (*How I Believe*, 1934, X, 110 E; 130 F)

ACTION *see also* **Energetics**
To act is to free from energy in order to create fuller being.

> In action I adhere to the creative power of God; I become coincidental with it; I become not only its instrument but also its living extension. (*The Divine Milieu*, 1926–7, 1932, IV, 34 E; 51 F)

ACTIVANCE *see also* **Activation**

> By "activance" I understood the capacity possessed by an intellectual or mystical perspective to develop and to superstimulate spiritual energies in us. (*Super-Humanity, Super-Christ, Super-Charity*, 1943, IX, 171 n.5 E; 216 n.1 F)

ACTIVANCE, PRINCIPLE OF THE MAXIMUM *see also* **Maximum, Principle of the**
Corollary of the principle of the maximum according to which the universe is to be seen as containing properties capable of maximizing the human power of action to its fullest.

> In relation to our zest for action, the world as a structure must offer a maximum power of excitation (an activance). To be simply actable, it must be supremely activating. (*Action and Activation*, 1945, IX, 175; 222–3 F)

ACTIVATION *see also* **Activance**
Capacity of an energy to absorb increased excitation. Activation is a capacity inherent in energy to increase its dynamism while activance is a capacity inherent in the animating cause of this energy (vision, Point Omega, divine person).

(a) To express completely the dynamic state of a living mass a formula having at least two terms must be used . . . the first measuring numerically the magnitude of thermodynamic dimensions, the second expressing the capacity of this energy to be used up, fairly quickly, in the direction of survival, multiplication or some super-arrangement of organized matter. Precisely, no doubt, because it is based on imponderables, this second term (that we shall call "activation") risks being neglected by bio-energetic experts. (*The Energy of Evolution*, 1953, VII, 365 E; 385–6 F)

(b) By activation I mean the special power possessed by certain factors (called activizing) to release through contact reserves of energy that, in the absence of such excitation, would remain dormant. (*The Singularities of the Human Species*, 1954, II, 260 E; 357 F)

ACTIVITIES *see also* **Passivities**
Everything charged with human energies. Opposite of passivities. These terms only appear in *The Divine Milieu*.

When we speak of "activity" we use the word in its popular sense without denying in any way and, indeed, affirming what takes place in the infra-experiential circles of the soul between grace and will. (*The Divine Milieu*, 1926–7, 1932, IV, 17 E; 29 F)

AGAPOSPHERE *see also* **Christosphere; Love; Theosphere**
Sphere of divine love. Personalist society where social relations are based on love (agape). For Teilhard, love is the sole moral imperative. Love is the root of all good – and of all evil (Henri de Lubac SJ, *The Eternal Feminine*, 38). Agape is the love that gives, as opposed to eros, the love that desires.

Love is the most universal, the most tremendous and the most mysterious of cosmic forces . . . Love is the primal and universal psychic energy . . . Love is a sacred reserve of energy . . . the blood of spiritual evolution. (*The Spirit of the Earth*, 1931, VI, 32–4 E; 40–2 F)

AGGREGATE *see also* **Aggregation; Complexity**
Accumulation, through inertia, of corpuscles and bodies that do not make up an organic totality. Any combination of unarranged elements (sand, stars, planets). Opposite of complexity.

The type of super-grouping, towards which the course of civilization is driving us, far from simply representing some material aggregates ("pseudo-complexes") where elementary freedoms are neutralized by the effects of large numbers or mechanized by geometric repetition, belongs on the contrary to the species of "eu-complex". (*Man's Place in Nature*, 1949, VIII, 101–2 E; 146–7 F)

AGGREGATION *see also* **Aggregate**

By complexity . . . I do not mean simple aggregation, that is, any combination of unarranged elements – such as a pile of sand – or even (allowing for a certain zonal classification due to gravity and irrespective of the multiplicity of substances going to make them up) the stars and planets. (*Man's Place in Nature*, 1949, VIII, 19–20 E; 28–9 F)

AHEAD *see also* **God-ahead**
Forwards movement of human beings towards the ahead. Progress of humanity by convergence.

And so, without our really appreciating it, an enormous psychological event is building up, at this very moment, in the noosphere: the meeting, no more and no less, of the above with the ahead; the confluence, on the christian axis, between the canalized stream of the old mysticisms and the newer but rapidly swelling torrent of the sense of evolution. Combined anticipation of a transcendent super-human and an immanent ultra-human: these two forms of faith illuminate and reinforce one another indefinitely. (*The Christian Phenomenon*, 1950, X, 206 E; 241 F)

ALL *see also* **Whole**
"All" and "Whole" are both translations of the Fr. "Tout" – a Teilhardian "keyword".

ALPHA (AND OMEGA) *see also* **Omega**
Symbol inspired by St. John: "*Ego sum Alpha et Omega*" (Rev 1:8).

(a) This eminently catholic notion of Christ. (*Note on the Universal Christ*, 1920, IX, 14 E; 39 F)

(b) In all honesty I am quite unable to envisage philosophically or mystically a universe without Christ "alpha and omega". (Letter to Henri Breuil, 25 March 1932, *Lettres inédites*, 183)

AMORIZATION *see also* **Love**
Activation of love within the framework of evolution. The infusion of activity with love. With valorization, one of the elements of moralization where amorization is a genuine communion of persons according to what is most interior to each of them. Teilhardian "keyword".

However, nothing less than the conjunction of Christ with Point Omega was necessary to my eyes being opened, in a flash, to the extraordinary phenomenon of a whole world aflame – through total amorization. (*The Heart of Matter*, 1950, XIII, 50 E; 62 F)

AMORIZE *see also* **Amorization**
Energize to the maximum.

(a) In the case of a universe unifying itself by a simple grouping of elements evolution is incontestably valorized – but this in a strictly impersonal atmosphere of collectivity. But in the case of a universe unifying itself under the influence of some already existing "supremely one", that is, by adhesion to a "supreme someone", it also becomes amorized. (*The Nature of Point Omega*, 1954, II, 272 E; 373 F)

(b) What is an amorized universe if it not a universe excited and activated to the very limits of its vital powers? (*From Cosmos to Cosmogenesis*, 1951, VII, 266 E; 274 F; see also *The God of Evolution*, 1953, X, 243 E; 291 F)

ANALOGY *see also* **Transformation, Creative**
Scholastic term.

1 In a static dimension, the passage from one level of being to another, respecting a certain homology while avoiding any confusion (traditional sense).

2 In an evolutionary dimension, in the world of phenomena, the

relationship between two successive stages of temporal progression, indissolubly uniting continuity and discontinuity (teilhardian sense).

(a) The whole difficulty lies in making "corrections by analogy". This is the old scholastic truism – rejuvenated within a temporal framework. (Letter to Henri de Lubac SJ, 29 April 1934, *Lettres intimes*, 274, 276–7)

(b) In an evolutionary universe it seem to me you can push the theory of analogy further than in an immobile world structure. (Letter to Henri de Lubac SJ, 29 October 1949, *Lettres intimes*, 381; see also XII, 83 E; 96 F)

ANALYSIS *see also* **Synthesis**

Regressive process by which the real is constantly broken down into ever simpler elements. Analysis highlights both partial structures and elements already constituted from the real but leaves the emergence of novelty and complete structures untouched.

The first step taken by the mind that wants to know what something is made of is to take it apart, to analyze it. The whole of science is based on this instinctive act. Science is essentially an analysis. Its methods of research, its conclusions, are governed by the principle that the secret of things lies in their elements, so that to understand the world it is sufficient to arrive at the simplest of terms from which it emerged. (*Science and Christ*, 1921, IX, 24 E; 50 F)

BL ANGELA DI FOLIGNO *see also* **Nicholas of Cusa**

Angela di Foligno (1248–1309) was one of the great Franciscan tertiary mystics of her age. Mother of several children, she became a Franciscan tertiary hermit after her husband's death. St. Francis was her model. She was known to her followers as the Mistress of Theologians. She has continued to have an important influence on mystics like Teilhard down to our own day (see also *Cosmic Life*, 1916, XII, 60 E; 49 F).

The eyes of my soul were opened and I discerned the fullness of God, in which I understood the whole world, here and beyond the sea, the abyss, the ocean and all things. In all these things I could see nothing but the divine power, in a way that was wholly indescribable. My soul was brimming over with wonder and I cried out, "The whole universe is full of God". (Angela di Foligno, *Le Livre des visions et instructions*, 1991, 73; see also *The Divine Milieu*, 1926–7, 1932, IV, 104 E; 140 F)

ANGUISH *see also* **Fear, Existential**

One of the stages of teilhardian optimism. Awareness of the disproportion between the person and what is capable of destroying the person: the cosmos in accelerated evolution and pressure of a collective totalization.

By the expression "existential fear" I do not mean the simple fear accidentally experienced by particularly timid people faced with the material or social risks that their existence holds for them. But taking these words in a more general and much deeper sense I use them here to represent anguish, not so much "metaphysical", as one says, but "cosmic" and biological, likely to affect everyone wise enough – or foolish enough – to try to find and measure the unfavorable depths of the world about them. (*A Phenomenon of Counter-Evolution in Human Biology*, 1949, VII, 183 E; 189 F)

ANTHROPIZATION *see also* **Hominization**

Accentuation of human somatic characteristics. Limited to a quantitative accumulation of traits that prepare the qualitative emergence of the human (hominization).

The hominids are placed historically and morphologically, in both cases, at the term of a long series of speciation (or, one could say, of a vast population of species) forming statistically a line from the eocene to the pliocene – the drift taking place along a principal and median axis of growing "anthropization" (globulization of the skull, flattening of the face, development of the hands, increase in height, etc.). (*The Law of Irreversibility in Evolution*, 1923, III, 251 E; 357 F)

ANTHROPOCENTRISM *see also* **Anthropomorphism; Movement, Neo-anthropocentrism of; Position, Anthropocentrism of**

Human species considered as center of the universe. Concept frequently criticized by Teilhard who insists on an anthropocentrism (or neo-anthropocentrism) grounded in evolution.

(a) Studied narrowly and apart from everything else by anthropologists and legal minds, the human being is a trial, even insignificant thing. Human individuality, too pronounced, masks the totality from our sight, so that as we consider the human our mind tends to fragment nature and to forget the depths of its connections and the boundless horizons it has: entirely the wrong kind of anthropocentrism. (*The Human Phenomenon*, 1938–40, I, 6 E; 30 F)

(b) Christianity is commonly criticized for its outdated reliance on both an anthropomorphism (of God) and an anthropocentrism (of human beings) . . . these too summary representations contain a lasting germ of truth. If, however, we see God, no longer as an ordinary center of consciousness (of the human type), but as a center of centers, and if we see human beings, no longer as the center of the world, but as an axis (or a spearhead) showing us the direction of the advance of the world (towards greater consciousness and personality), then we can avoid the weaknesses of anthropomorphism and anthropocentrism and yet retain everything required by christian dogma. And this simply by an enriching change of dimension. (*Some General Views on the Essence of Christianity*, 1939, X, 137 E; 159 F)

ANTHROPODYNAMICS *see also* **Energetics; Noodynamics; Sociodynamics**

Branch of anthropology that defines an energetics fully centered on the human.

> We need a generalized physics or energetics capable of integrating an anthropodynamics and an anthropogenesis. (*The Phenomenon of Man*, 1954, XIII, 155 E; 199 F)

ANTHROPOGENESIS *see also* **Ethics; Hominization**

Appearance and development of the human group by crossing a specific threshold (the breakthrough to reflection) that corresponds to a higher state of cosmic arrangement (continuity) and a change of nature (discontinuity). The genesis of humanity will be the object of a future synthetic science covering the formation and, above all, the future of human beings, capable of harmonizing the different scientific disciplines – physical, biological, anthropological, moral, etc.

(a) Certainly, because of the great mass of factors that condition it, anthropogenesis depends on sidereal, planetary and biospherical energies whose working will always escape us. But, because, of its most axial and most active germ – because of the progress of the (individual and collective) nervous system – is it not on the point of falling under the extended beam of our power of invention? (*The Phyletic Structure of the Human Group*, 1951, II, 164 E; 226 F)

(b) Non-biologists often forget that underlying the various rules of ethics, economics and politics there are written into the structure of our universe certain general and absolute conditions of organic growth. To determine in the case of the human these basic conditions of biological progress should be the specific field of the new anthropology, the science of anthropogenesis, the science of further human development. (*Abstract of Paper on the Significance and Trend of Human Socialization*, 9 April 1948 (Or. Eng.), Claude Cuénot, *Teilhard de Chardin*, 283 E; 342–3 F; see also letter to Henri de Lubac SJ, 9 November 1948, *Lettres intimes*, 377)

ANTHROPOGENETICS

Discipline concerning the overall conditions of human evolution. Rare term.

> New science: anthropogenetics = science of general determinisms in human evolution. Function: to determine the direction and steps of all eugenics, research and maintenance. (See also Journal, 25 March 1948, in Claude Cuénot, *Nouveau Lexique*, 44)

ANTHROPOMONISM

Doctrine, questioned by Teilhard, according to which there is only one humanity and, therefore, only one incarnation.

Here we have the essence of the difficulty of "christianity = anthropocentrism" . . . This difficulty is basically found in "anthropomonism" (and, more particularly, in mariology) and the problem of considering how to accept the christic self having to suffer again on some future planet . . . Once again, how can we separate christology (and mariology) from anthropomonism? But, for all the world, I repeat, this is something we must do. (Retreat, Peiping, 20–5 October 1945, 5th day, in Claude Cuénot, *Nouveau Lexique*, 44)

ANTHROPOMORPHISM *see also* **Anthropocentrism**

Application of human form and attributes to God. Concept rejected by Teilhard.

(a) What gives christianity its particular effectiveness and tonality is the fundamental idea that the supreme focal point of unity is not only reflected in every element of consciousness that it attracts but, in order to bring about final unification, has had to "materialize" itself as an element of consciousness (the historic, christic "I"). In order to act effectively, the center of centers reflected itself on the world as a center (= Jesus). This concept of Christ, not only as prophet and man exceptionally conscious of God, but also as "divine spark", shocks the "modern" mind at first glance as an outmoded anthropomorphism. But we should note:
(i) that the modern reaction against anthropomorphism has gone much too far, to the point of making us doubt a divine ultra-personality . . .
(ii) and that, psychologically, the astonishing power of mystical development displayed by christianity is indissolubly linked to the idea that Christ is historical. Suppress this key element and christianity becomes but one of many "philosophies": it loses all its force and vitality. (*Some General Views on the Essence of Christianity*, 1939, X, 136 E; 157–8 F)

(b) We must . . . recognize that, as a result of a certain excess of anthropomorphism (or primitive nationalism), the judæo-christian mystical current has had some difficulty in breaking away from a point of view where Oneness was too exclusively sought in singularity rather than the synthetic power of God. God loved above all things (rather than in or through all things). Hence a certain "thinness" in the mysticism of many prophets and saints, mysticism too "Jewish" or too "human" in a narrow sense – insufficiently cosmic and universalist (there are exceptions, of course: Eckhart, Francis of Assisi, St. John of the Cross . . .). (*Some Notes on the Mystical Sense*, 1951, XI, 210–1 E; 228 F)

ANTHROPOSPHERE *see also* **Noosphere**

First name for the noosphere.

(a) *Credo unitati* (I believe in unity). Anthroposphere. (Notes and Sketches, Cahier 7, 17 January 1920, in Claude Cuénot, *Nouveau Lexique*, 44)

(b) Who will be the Suess of the anthroposphere? (Notes and Sketches, Cahier 7, 25 February 1920, in Claude Cuénot, *Nouveau Lexique*, 44)

ANTI-ENTROPY *see also* **Neguentropy**

ANTI-TRANSFORMISM *see also* **Fixism**

APOLOGETICS

1 In a general sense, dialectical approach that seeks to demonstrate the probability and reasonableness of christian faith.
2 In a teilhardian sense, second of three stages of teilhardian thought (physics, apologetics, mysticism). Synonym of dialectics.

(a) I have not forgotten the project of an "apologetics" . . . but the more I think about it the more this essay takes in my mind the form of a simple statement of "my personal reasons for believing". (Letter to Bruno de Solages OCarm, 28 August 1934, *Lettres intimes*, 293)

(b) If you have the patience to read the pages I sent (at the beginning of January) to Le Roy entitled "How I Believe", you will see the entire structure of my faith is built upon the basis of a complete adhesion to the supreme value of what is developing around us, and in us, in the universe. And this is no fiction: in reality, the whole of my religion depends upon this initial belief in the world and in the future of the world. (Letter to Christophe Gaudefroy, 14 February 1935, *Lettres inédites*, 106–7)

(c) Essentially, the thought of Fr. Teilhard de Chardin is expressed not in a metaphysics but in a sort of phenomenology . . . In a second phase, Fr. Teilhard constructs . . . an apologetics: under the illuminating influence of grace our minds recognize in the underlying properties of the christian phenomenon a manifestation (reflection) of omega on human consciousness and identify the omega of reason with the Universal Christ of revelation. (*My Intellectual Position*, 1948, XIII, 143–4 E; 173–4 F)

(d) In order to avoid any future misunderstanding I think it would be useful if I were to present here clearly defined the successive stages of my apologetics or, if you prefer, my dialectics:

1st stage: the human phenomenon and the existence of a transcendent God . . .
2nd stage: evolutionary creation and the expectation of a revelation . . .
3rd stage: the christian phenomenon and faith in the incarnation . . .
4th stage: the living Church and Christ-Omega. (*Outline of a Dialectic of Spirit*, 1946, VII, 144–8 E; 149–55 F)

APOSTOLATE, TEILHARD'S

(a) I, Lord, for my very lowly part, would wish to be the apostle and (if I dare be so bold) the evangelist of your Christ in the universe. (*The Priest*, 1918, XII, 219 E; 329 F)

(b) I am a pilgrim of the future on my way back from a journey made entirely in the past. (Note, October 1923, *Letters from a Traveler*, 60; *Lettres de voyage*, 105)

(c) I am preparing a sort of small "summa", simpler . . . but more structured, of the essence of my ideas. This will be entitled "The Spiritual Phenomenon" (three parts: Spiritualization, Personalization, Moralization). The third part will interest you. I am gradually coming round to the idea that "love of God" is a specifically christian phenomenon, the specific reaction of the personal term of the universe on the "noosphere". (Letter to Bruno de Solages OCarm, 3 February 1937, *Lettres intimes*, 328)

(d) I do not believe that I have ever been more sincerely convinced that there is no possible way out for the "noosphere" outside the christian axis. (Letter to Christophe Gaudefroy, 20 June 1929, *Lettres inédites*, 77)

(e) What I am putting before you are suggestions rather than affirmations. My principal objective is not to convert you to ideas that are still fluid but to open horizons for you, to make you think. (*Man's Place in the Universe*, 1942, III, 217 E; 306 F)

(f) If I have had a mission to fulfil, it will only be possible to judge whether I have accomplished it by the extent to which others go beyond me. (Conversation with Max Henri Bégouën, summer, 1954, *Cahiers Pierre Teilhard de Chardin*, N° 2, 1960, 35)

AQUINAS, ST. THOMAS *see also* **Fathers, Greek; St. Thomas Aquinas**

ARCHETYPES, STRUCTURAL *see also* **Analysis; Convergence; Corpusculization; Dispersion; Divergence; Groping; Moleculization; Schemes, General; Structure, Scaled; Synthesis; Union**

Fundamental framework of reference to which the real is obedient and which makes it thinkable.

Structural archetypes: inertia; union; divergence, convergence, groping. (See also *Note on the Possible Preparation of a Lexicon*, 1951, in Claude Cuénot, *Nouveau Lexique*, 45)

ARRANGEMENT *see also* **Curve**

Improbable effect of synthesis and complexification through hierarchic organization and centration.

1 In a broad sense, mediation between the tangential and the radial.

What . . . is the particular function linking both the radial and tangential energies of the world together experimentally in their respective developments? Clearly, it is arrangement: whose successive progressions are interiorly lined, as we can observe, with a continuous growth and deepening of consciousness. (*The Human Phenomenon*, 1938–40, I, 92 E; 155 F)

2 In a more specific sense, one of the two forms of tangential (physical) energy where the radial (psychic or spiritual) assumes a dominant value through opposition to radiation, the more rudimentary form of tangential energy where the radial has a negligible value.

It is probably necessary to distinguish two sorts of tangential energy: one of radiation (the maxima for very small radial values – as in the case of the atom); the other of arrangement (perceptible only for higher radial values – in the case of living beings and the human). (*The Human Phenomenon*, 1938–40, I, 31 n. E; 62 n.1 F)

ARRANGEMENT, CURVE OF

Metaphor borrowed from non-euclidean geometry. One of the fundamental drifts of the universe that pushes matter, in favorable circumstances, towards increasing complexification.

No longer universal "attraction" gradually becoming united with the cosmic mass but the still unperceived and unnamed power compelling matter (as it concentrates under pressure) to become arranged in increasingly larger, more differentiated and more organized corpuscles. Over and beyond the curve of union, the curve of arrangement. (*The Heart of Matter*, 1950, XIII, 33 E; 43 F)

ARTIFICIAL

Reflective nature. Relatively complex organs and tools created by human intelligence and detachable from the biological organism. In one sense, an extension of nature, in another, its opposite.

Because we have assumed in principle that the artificial contains nothing natural (that is, because we have failed to appreciate that the artificial is derived from the humanized natural) we have failed to recognize such clear vital analogies as the bird and the plane or the fish and the submarine. (*Hominization*, 1923, III, 59 E; 87 F)

ASCENT *see also* Descent

Moral dimension. Antonym of descent.

Properly speaking, there are neither sacred nor profane things, neither pure nor impure things. There is only a good direction and a bad direction: the direction of ascent, of expanding unification, of greater spiritual effort; and the direction of descent, of constricting egoism, of materializing enjoyment. (*The Evolution of Chastity*, 1934, XI, 72 E; 78 F)

ASCESIS *see also* Detachment

Meditation. Practice of the virtues through will, privation, etc.

(a) Neither of the two components of interior life (natural growth and then diminishment in Christo) can or should eliminate the other. There are not two opposed asceses, one of development, the other of mortification. But there are two phases able to come together in a flexible and movable equilibrium. (*Forma Christi*, 1918, XII, 263 E; 380 F)

(b) Hitherto, ascesis tended to reject: to become holy it was necessary, above all, to be deprived. Henceforth, by virtue of the new moral aspect assumed by matter in our eyes, spiritual detachment will take the form of a conquest. (*The Evolution of Chastity*, 1934, XI, 72 E; 79 F)

ASSUMPTION, DOGMA OF THE *see also* **Church, Infallibility of the**
Apostolic Constitution *Munificentissimus Deus* (1950).

To define the Assumption is to affirm (through the facts) that "revelation continues", that is, that dogma grows (that it lives): no more magnificent proof that christianity (like the living and the human) evolves (in the correct "genetic" sense of the word). (Letter to Jeanne-Marie Mortier, 25 August 1950, *Lettres à Jeanne Mortier*, 69)

ASTRONOMY

I do not understand why astronomy is still thought of and even practiced as a conservative science. It is far more revolutionary than all the historical sciences of life. (Letter to Auguste Valensin SJ, 28 February 1920, *Lettres intimes*, 56)

ATTRACTION *see also* **Pull**
Energy of universal attraction on the level of persons. Simultaneously in continuity and discontinuity with the cosmic process.

Force, in its old form, only expresses the power of the human over the extra- or infra-human. At the heart of humanity, among humans, it has been changed into its spiritual equivalent – energy of attraction instead of repulsion. (*The Moment of Choice*, 1939, VII, 18 E; 24 F)

AUGUSTINE, ST. *see also* **Fathers, Greek**
Aurelius Augustinus (354–430), bishop and doctor of the Church. Augustine was the first to introduce the christocentric theme into human history. Teilhard does so with cosmic history.

(a) Was it not St. Thomas who, comparing (what today we should call the fixist) perspective of Latin Fathers like St. Gregory to the "evolutionary" perspective of the Greek Fathers and St. Augustine, said of the latter, "Magis placet" (II Sent, d.12; g.I.a.I)? Let us be glad to strengthen our minds by contact with this great thought! (*What Should We Think of Transformism?*, III, 1930, 154 E; 217 F)

Magis placet = it is more pleasing

(b) In the presence of St. Augustine, St. Mary Magdalene or St. Lydwina, no one hesitates to think of *felix dolor* or *felix culpa*. With the result that, up to this point, we continue to "understand" Providence. (*The Divine Milieu*, 1926–7, 1932, IV, 67 E; 91 F)

AUTO-

Prefix forming reflective compounds expressing a dynamism that is immanent in the world, e.g. auto-accelerated, auto-adoration, auto-arrangement, auto-birth, auto-centration, auto-center, auto-centric, auto-centrism, auto-cerebralization, autochton, auto-convergence, auto-critic, auto-destruction, auto-direction, auto-evolution, auto-evolutionary, auto-evolver, auto-extend, auto-fecundation, auto-fertility, automatism, auto-organization, auto-rebound, auto-regulation, auto-selection, auto-sufficiency, auto-suggestion, auto-transformation, autotroph, auto-unification . . .

AUTO-CEREBRALIZATION *see also* **Cerebralization**

Concerted rebound of research on the intelligence from which it proceeds that seeks to modify the thinking organ through its own neurological techniques.

> The revolutionary possibility, made feasible by the operation of social innervation, of a concerted wave of research into the intelligence from which it proceeds, becomes apparent: collective cerebralization (in a convergent milieu) using the fine point of its enormous power to complete and improve anatomically the brain of each individual . . . cerebralization (higher effect and parameter of cosmic evolution) closing in on itself in a process of self-completion; an auto-cerebralization of humanity becoming the most concerted expression of the reflective rebound of evolution. (*Man's Place in Nature*, 1949, VIII, 110–11 E; 158–60 F)

AUTO-EVOLUTION *see also* **Self-evolution**

Concerted rebound of thought on evolution (of which it is a phenomenal product) with the acquisition of control over its own underlying energies.

> The zest, the will to act, would be reborn and would rebound in our hearts in line with the ever greater evolutionary effort we must make in order to ensure the progress of a complexity that is ever heavier to carry. This, we must not forget, is the dynamic condition essential to the survival of a biogenesis that has definitely entered into us from a state of evolution to a state of auto- or self-evolution. (*Transformation and Continuation in Man of the Mechanism of Evolution*, 1951, VII, 309 E; 323 F)

AXIAL *see also* **Axis; Radial**

(a) For me, the Church is the axial current of life . . . Small as it appears besides the secular cycles of the religious spirit . . . it is the actual depository of life, even if it still does not understand all the richness of the mystery and force of the ferment of which it is the vehicle. (*Journal*, 13 December 1918, *Journal*, 377)

(b) By virtue of the "law of complexity-consciousness", that has always guided

us till now, one can say there exists, in every corpuscle, two levels of operation: one (the tangential) binding physico-chemically, that is, through complexity, the particular corpuscle in relation to other corpuscles in the universe; the other (the radial or axial) leading directly from consciousness to consciousness and manifesting itself, at the level of humanity, in the different psychological phenomena . . . of unanimity and co-reflection. (*The Singularities of the Human Species*, 1954, II, 265 E; 362–3 F)

AXIS *see also* **Axial**

In an evolutionary dimension, function analogous to essence in a static dimension. In a teilhardian perspective, the real is energy and dynamism, hence oriented movement. In the old perspective, it is based on the immutability of the idea or essence.

(a) It is the cosmic axis, of both physical arrangement and psychic interiorization, revealed by the basic drift or orthogenesis, to which I shall constantly refer, every time it is necessary to evaluate the significance of an event or process as an absolute value. Axis of complexity-consciousness, I shall call it, usefully transposable . . . into axis of cephalization (or cerebration) following its appearance in the nature of nervous systems. (*The Singularities of the Human Species*, 1954, II, 215 E; 304 F)

(b) The special apostolate . . . that seeks to sanctify not only a nation or social group but the very axis of human drive towards spirit, contains two distinct phases: one, natural, providing an introduction to the christian faith; the other, supernatural, showing the (revealed) prolongations of earthly activity. (*Note on the Presentation of the Gospel in a New Age*, 1919, XIII, 214 E; XII, 404 F)

(c) More than ever I find myself bound to the axis of the Church. (Retreat notes, 11.1939, in *Lettres intimes*, 338)

BAPTISM

One of seven sacraments recognized by the Catholic and Orthodox Churches. Symbol of purification and renewal.

> The full mystery of baptism is no longer to cleanse but (as the Greek Fathers understood so well) to plunge into the fire of the purifying struggle "for being". No longer the shadow, but the passion of the Cross. (*Christology and Evolution*, 1933, X, 85 E; 103–4 F)

BEING *see also* Dialectics; Non-being

Existence, substance, essence. Fundamental movement or passage from a static to a dynamic dimension. Being is union, that is, on the level of creation, action is union, on the level of participated being, the process of union (see also *My Fundamental Vision*, 1948, XI, 193 n.27 E; 208 n.1 F).

> In classical metaphysics it is usual to deduce the world from the notion of being, considered irreducibly original. Backed by recent research in physics that has proved (contrary to "popular" belief underlying the whole of *philosophia perennis*) that movement is not independent of the moving body, rather that the moving body is physically engendered (or, more exactly, co-engendered) by the movement that inspires it, I shall try to show that a dialectics richer and suppler than others is possible, if from the outset, being, far from representing a terminal and solitary notion, is seen in reality as definable (at least genetically, if not ontologically) by a particular movement indissolubly associated with it – that of union. (*My Fundamental Vision*, No. 26, 1948, XI, 192–3 E; 207–8 F)

BEING, BETTER *see also* Being, Fuller

Arrangement of the real for human well-being. Opposite of fuller being to the extent that it lacks any ontological growth.

> The "social" demands of worker priests for better being mask aspirations for, and neo-humanistic faith in, fuller being. (*Research, Work and Worship*, 1955, IX, 218 n.1 E; 287–8 n.1 F)

BEING, FULLER *see also* Being, Better

Appearance, at each stage of evolution, of a new reality that constitutes a surplus of consciousness and, consequently, an ontological enrichment.

Opposite of better being.

> We . . . do not place fuller being in the realm of comfort or virtue but in the growing control of the world through thought (that is, in a growing force for good or for evil) . . . We believe in progress. (*My Universe*, 1924, IX, 81 E; 110 F)

BEING, PARTICIPATED *see also* **Being**

Created being that, while acquiring increasing autonomy, cannot exist without the support of the creative will of God. With absolute being, forms one of the two poles of being. Thomist expression.

> Since God can only be conceived as monopolizing in himself the totality of being, either the world is no more than an appearance or it is a part, an aspect or a phase of God in itself. In order to escape the dilemma the christian metaphysician has developed the idea of "participated being", a lower or secondary form of being ("sub-being" we might say), freely drawn from "non-being" by a special act of transcendent causality, "*creatio ex nihilo*". (*Action and Activation*, 1945, IX, 180 E; 227 F)

BIOGENESIS

Second phase of evolution. Evolution of life. Threshold of emergence of the process of complexification that reveals the principal axis of cosmogenesis and leads to the growth of increasingly self-centered autonomous creatures.

> "Compression/Complexity/Consciousness"
> or again, if you prefer,
> "Compression/Competition/Complexity/Consciousness".
> Such, in reality, is the three (or four) term formula really capable of expressing the process of biogenesis along its whole chain. (*The Phyletic Structure of the Human Group*, 1951, II, 159 E; 219–20 F)

BIOLOGY

Science of physical life that deals with organized beings or animals and plants, their morphology, physiology, origin and distribution.

> Everything changes if (as suggested by my curve of corpusculization) life is nothing other, scientifically speaking, than a specific effect (the specific effect) of complexified matter . . . From the point of view of biology nothing other than the physics of the very highly complex – it is interesting to note how everything falls into place in our experience. (*Man's Place in Nature*, 1949, VIII, 24 E; 34–5 F)

BIOSPHERE

Sphere of life. Term first used by Eduard Suess (1831–1914). Teilhard uses it to express the organic (non-thinking) layer of the earth between the inorganic layers (lithosphere and hydrosphere) and the psycho-spiritual or thinking layer (noosphere).

(a) By biosphere we understand, not, as some mistakenly suppose, the peripheral zone of the world to which life is confined, but the actual pellicle of organic substance we see enveloping the earth today – truly a structural layer of the planet, despite its thinness. (*Man's Place in Nature*, 1949, VIII, 40 E; 57 F)

(b) The earth's thinking envelope (as well as its merely living envelope) is not simply an aggregate or moral unity but an organic whole *sui generis*. (*The Phenomenon of Man*, 1928, IX, 91 E; 122–3 F)

BIOTA *see also* **Layer**

Evolutionary unity defined by its common origins and living solidarity.

From the evolutionary (we could even say "physiological") perspective, taken as a unit the placental mammals constitute what I shall call here conventionally a biota. By this I mean a grouping of verticils whose elements are not only related by birth, but also support and mutually complete each other in the effort to stay alive and to propagate. (*The Human Phenomenon*, 1938–40, I, 76 E; 133 F)

BRAIN, NOOSPHERIC

Individuals . . . assert themselves, through their phren, as constituent elements of the "noospheric brain" (the organ of collective human thought). (*Man's Place in Nature*, 1949, VIII, 111 n.1 E; 160 n.1 F)

BRANCH

Highest division in the living world.

(a) In the main (vertebrate) branch, we now hold the vastest type of assemblage that systematics has yet recognized and defined within the biosphere. In addition to the vertebrates, there are two, and only two, other main branches that contribute to form the main branching structure of life: the worm and arthropod branch and then the plant branch. (*The Human Phenomenon*, 1938–40, I, 83 E; 142 F)

(b) Through our efforts of analysis, life peels apart. It disarticulates indefinitely into an anatomically and physiologically coherent system of overlapping fans. Of faintly sketched micro-fans of sub-species and races. Of already broader fans of species and genera. Of more and more enormous fans of biotas, and then of layers, and then of branches. (*The Human Phenomenon*, 1938–40, I, 89 E; 150 F)

BREAKTHROUGH *see also* **Threshold**

For those who know how to see an essential complement to this great cosmic event of reflection in the discovery of the shape of what we might call the "breakthrough to amorization". Even after the flash of individual self-revelation, primitive humans would have remained incomplete if, on meeting the opposite sex, they had not burst into flame in a centric person to-person attraction completing the appearance of a reflective monad, the formation of an affective monad. (*The Heart of Matter*, 1950, XIII, 60 E; 73–4 F)

CANALIZATION *see also* **Canalization, Mutation by**

We could say that mutations in a given group of living creatures can act (and thereby show themselves) in at least three ways: by dispersion, by radiation, by canalization . . . By canalization . . . when the very strongly "polarized" changes of form converge and come together in a common direction. (*The Movements of Life*, 1928, III, 145–6 E; 203–4 F)

CANALIZATION, MUTATION BY

Additive change that, linked with preceding and succeeding changes, represents an evolutionary process obedient to a pattern of convergence, e.g. horses. Teilhard integrates mutation with orthogenesis.

Psychic simplicity, as we know it, is born of the multitude. It flourishes on organic complication, sovereignty numerous and sovereignty conquered. Proportional to the surmounted multitude that it shelters, it corresponds inexorably to the maximum tendency in being to become decomposed. (*The Struggle against the Multitude*, 1917, XII, 96–7 E; 133–4 F)

CATHOLICISM *see also* **Church; Church, Catholic**

Community of local churches and their bishops in communion with the Church of Rome and its bishop, successor of St. Peter and St. Paul.

(a) I must insist on the fact that . . . the interior attitude I have just described has the direct effect of linking me ever more inescapably to three convictions which form the marrow of christianity: the unique value of the human as the spearhead of life; the axial position of catholicism in the convergent fascicle of human activities; and finally the essential consummating function assumed by the risen Christ at the centre and summit of creation . . . I now feel myself more indissolubly bound to the hierarchical Church and to the Christ of the Gospels than ever before in my life. Never has Christ seemed to me more real, more personal, more immense. (Letter to Jean-Baptiste Janssens SJ, 12 October 1951, *Lettres intimes*, 1974, 399–400; *Lettres familières*, 115; *Letters from My Friend*, 105)

(b) Christianity, by its very essence, is much more than a fixed system, formulated once and for all, of truths that must be accepted and preserved literally. For all its basis in a nucleus of "revelation", it represents, in fact, a spiritual attitude in process of continual development: development of a christic consciousness that keeps pace with and is required by the growing consciousness of humanity. Biologically, it behaves like a "phylum". By biological necessity, therefore, it must have the structure of a phylum, that is, it must constitute a progressive and coherent system of collectively associated spiritual elements. Clearly, hic et nunc, nothing within christianity but catholicism possesses these characteristics.

There are doubtless many individuals outside catholicism who love and discern Christ and who are united to him as closely (and sometimes even more closely) than catholics. But these individuals are not grouped together in the "cephalized" unity of a body that reacts vitally, as an organic whole, to the combined forces of Christ and humanity. They are fed by the sap that flows in the trunk without sharing in its youthful surge and elaboration at the very heart of the tree. Experience shows that, not only logically, but factually, it is only in catholicism that new dogmas continue to germinate and, more generally, new attitudes are formed that, by process of continued synthesis between ancient Credo and views newly-emerged in human consciousness, pave the way for the coming of a christian humanism.

Evidence suggests that if christianity is indeed destined, as it professes and feels, to be the religion of tomorrow, it is only through the organized and living axis of its Roman Catholicism that it can hope to match and assimilate the great modern currents of humanitarianism. To be catholic is the only way of being fully and completely christian. (*Introduction to the Christian Life*, 1944, X, 167–8 E; 196–7 F)

CENTER *see also* **Centration**

One of four fundamental symbols, with the circle, cross and square. Frequently used by Teilhard as dynamic focal point of union appearing at a particular stage of evolution (life) and open to continuously increasing self-concentration. Not a static point in space. Omega is the supreme center.

(a) At every degree of size and complexity, the cosmic corpuscles or grains are not only, as physics recognizes, centers of universal dynamic radiation but, somewhat like humans, they possess and represent . . . a small "inside" (however diffuse or fragmentary it may be . . .) in which is reflected, in more or less rudimentary form, a particular representation of the world: psychic centers in relation to their own selves and, at the same time, infinitesimal psychic centers of the universe. (*Centrology*, 1944, VII, 101 E; 106–7 F)

(b) It is precisely because God is the center that he fills the whole sphere. (*The Divine Milieu*, 1926–7, 1932, IV, 102 E; 136 F)

CENTER, SUPER-PERSONAL *see also* **Super-center**

Universal center of human and cosmic evolution.

A real pole of psychic convergence: a center different from all other centers that it "super-centers" by assimilating them; a person distinct from all other persons whom it fulfils by uniting them to itself. The world would not function if there did not exist somewhere ahead in time and space "a cosmic point Omega" of total synthesis. (*Human Energy*, 1937, VI, 145 E; 180 F)

CENTRATION *see also* **Decentration; Excentration; Interiorization; Super-centration**

1 General process of being that is folded back on itself, interiorized and unified.

> Each one (of the first centered corpuscles) must have emerged in the biosphere, on its own account, as a result of the repeated play of large numbers; each one, in order to pass from pre-life to life, must have crossed a particular critical point of centration (the closing on itself of a chain of segments) whose replica in higher form we shall meet later in the case of reflection. (*Centrology*, 1944, VII, 106 E; 112 F)

2 First stage of the existential dialectic of happiness and of all spiritual life (unification of the self within itself constituting the first condition for the gift of the self and, thereby, a higher unification). Conquest of personal autonomy.

> If we are to be fully ourselves and fully alive, we must:
> (1) be centered on ourselves;
> (2) be decentered on the "other";
> (3) be super-centered on one greater than ourselves . . .
> Not only physically, but intellectually and morally, we are only human if we develop ourselves. (*Reflections on Happiness*, 1943, XI, 117 E; 129–30 F)

CENTRIC *see also* **Christic**

One of two evolutionary process: cosmic (natural) and christic (supernatural). Milieu of spiritual unity where the christic assimilates and harmonizes both the cosmic and the human in preparation for the parousia.

> A remarkable and astonishing region, indeed, where through the meeting of the cosmic, the human and the christic, a new area, the centric, opens up, where the multiple oppositions that go to make up the unhappiness or the anxieties of our existence tend to disappear . . . A miraculous specific effect of the centric that neither dissolves nor subjugates the elements it brings together but personalizes them: precisely because its way of absorbing them is always to "centrify" them more and more. (*The Heart of Matter*, 1950, XIII, 49–50 E; 61–2 F)

CENTRICITY *see also* **Centration; Interiorization**

Simple qualitative synonym of centration or interiorization

> If we look carefully at things, we see the "centricity" of an object does not correspond, in the world, either to an abstract quality or to some sort of "all or nothing" that knows neither nuance nor degree. But it represents, on the contrary, an essentially variable magnitude, proportional to the number of elements and connections in each cosmic particle. A center is simpler and deeper the greater its density and radius at the heart of a sphere. (*The Atomism of Spirit*, 1941, VII, 31 E; 37–8 F)

CENTRICITY, FRAGMENTARY *see also* **Centrogenesis; Pre-life**
First or pre-centric stage of centrogenesis. State of matter that does not yet have a real inside but only "elements of immanence" or a simple disposition towards psychism. Characteristic of the inorganic and pre-life.

> In the case of "inanimate" matter . . . we can picture figuratively to ourselves the cosmic nuclei (molecules, atoms, electrons . . .) as incompletely closed on themselves: elements that must already possess a sort of psychic curve (otherwise they would not exist) in the form of fragments open at either end like the broken segments of a sphere or a circle. (*Centrology*, 1944, VII, 105 E; 111 F)

CENTRICITY, PHYLETIC *see also* **Biosphere; Centrogenesis; Phylum**
Second stage of centrogenesis. Differs from fragmentary centricity by greater centration and by belonging to a phyletic line (phylum). Unlike the eu-centric, has not yet crossed the threshold of perfect centration and personal autonomy. Characteristic of the biosphere.

> Quite different is the behavior of closed corpuscles, the specific elements of the biosphere. Although barely centered on itself, such a corpuscle shows itself possessed of a remarkable power of self-complexification (and, therefore, of autocentration) . . . In other words, driven by a sort of lifting power, it tends to raise itself up radially, like a rocket, from the mono-cellular to the poly-cellular, in the general direction of omega, tracing a phylum. (*Centrology*, 1944, VII, 106–7 E; 112–13 F)

CENTRO-COMPLEXITY *see also* **Arrangement; Complexification; Consciousness**
Coefficient measuring the degree of centricity and complexity where centricity (interiorization) is a function of complexity. Coefficient measuring the degree of psychism or "consciousness" (in a teilhardian sense).

> The most essential, the most significant characteristic of any group of unities making up the universe is distinguished by a certain degree of interiority, that is, of centricity (soul), itself a function of a certain degree of complexity (body and, more especially, brain). This coefficient of centro-complexity (or, that amounts to the same thing, consciousness) is the true absolute measure of being in the beings that surround us. (*Centrology*, 1944, VII, 102 E; 107 F)

CENTROGENESIS

1 Three-stage process of convergence or centration:

(1) fragmentary centricity (or pre-centricity);
(2) phyletic centricity;
(3) eu-centricity.

Convergence of the universe along its axis of centro-complexity. Creative union.

2 The existence and characteristics of a prime mover ahead, omega, are demonstrated by the need of centrogenesis for love and immortality. Its success is dependent on the future being open to someone already present, someone in whom humanity can attain, without dissolution, its peak of centralization and personalization.

By virtue of the genetic relationship (centrogenesis) that causes centricity (unity) to depend on complexity (multiple), the two aspects of the real, spiritual and material, necessarily and complementarily suit one another like the two faces of the same object or, better, like the two terms "*a quo*" and "*ad quem*" of the same movement. (*Centrology*, 1944, VII, 124–5 E; 131 F)

CENTROLOGY *see also* **Union, Dialectic of**

Science and synthetic vision of the process of centration.

While cosmic corpuscles in the static universe of monadology "have neither doors nor windows", from the evolutionary point of view of centrology, they are triply interdependent within the centrogenesis into which they are born. (*Centrology*, 1944, VII, 104 E; 109 F)

CEPHALIZATION *see also* **Cephalization, Law of**

Evolutionary tendency, particularly marked in vertebrates, of the ganglion cells (nervous system and sense organs) to concentrate in the head. In other words, increase in the mass of cerebralized matter, parallel to the increase in complexity of the cerebral connections.

The variation of the nervous system or, more precisely, the variation of its cephalized portion or, more simply and in one word, cephalization, is the master key we need. (*Man's Place in Nature*, 1949, VIII, 48 E; 69 F)

CEPHALIZATION, LAW OF *see also* **Orthogenesis, Basic**

Teilhardian law governing the continuously rising curve of life, with time as the abscissa and proportion (quality and quantity) as the ordinate of nervous matter ("precephalization") and, above a certain level, cerebralized matter (in the proper sense of cephalization) appearing on earth at each stage of evolution.

(a) Law of cephalization. Whatever the animal group (vertebrate or anthropoid) whose evolution we study, it is a remarkable fact that, in every case, the nervous system increases in time in both volume and arrangement and, simultaneously, concentrates in the frontal, cephalic region of the body. (*Super-Humanity, Super-Christ, Super-Charity*, 1943, IX, 155 E; 199 F)

(b) The law of cephalization itself is only the higher form assumed among living beings by the law of complexity. (*Super-Humanity, Super-Christ, Super-Charity*, 1943, IX, 156 E; 200 F)

(c) To localize, as catholics do, the permanent organ of phyletic infallibility in the councils, or, by an even more advanced concentration of christian consciousness, in the Pope (formulating and expressing, not his own ideas, but the teaching of the Church) fully conforms with the great law of "cephalization" that governs all biological evolution. (*Introduction to the Christian Life*, 1944, X, 153 E; 181 F)

CEREBRALIZATION *see also* **Auto-cerebralization; Cephalization**

This criterion . . . of cephalization or cerebralization (is) understood in the precise technical sense of the "development" of a neo-cortex or neo-pallium. (*Man's Place in Nature*, 1949, VIII, 49, 54 E; 70, 77 F)

CEREBRALIZATION, PARAMETER OF

Quantitative and qualitative rate of growth of the nervous system (complexification) that allows identification of the fundamental law of evolution.

First result obtained by application of the parameter of cerebralization: the principal axis of cosmic involution (or corpusculization) on earth passes through the mammal branch. (*Man's Place in Nature*, 1949, VIII, 49 E; 70 F)

CEREBRATION *see also* **Cephalization**

Chief phenomenon of anthropogenesis where, over time, the brain of the highest order of mammals, the primates, becomes increasingly elaborated and convoluted.

Whether insects or invertebrates, it is rare for a living group of any kind, provided we can follow it over a long enough space of time, not to show a notable advance in what we can call indifferently cephalization or cerebration. (*The Singularities of the Human Species*, 1954, II, 220 E; 309 F; see also diagram, *The Phyletic Structure of the Human Group*, 1951, II, 141 E; 197 F)

CHANCE *see also* **Large Numbers, Law of**

Fortuitous meeting of elementary determinisms, imitating the intentional final act but in reality subject to the law of large numbers and the calculation of probabilities. Important category of teilhardian thought used here in its universal sense.

Everywhere, lost to view before me, the future undulates like some imperceptibly treacherous milieu . . . The only fixed lines I can see are the inhuman characteristics of certain laws of probability. As chance arises from a group of assembled determinisms, so a secondary order of determinisms is born of the meeting of all chances. (*Operative Faith*, 1918, XII, 229 E; 340–1 F)

CHARITY, EVOLUTIONARY
Evangelical charity dynamized by the new perspectives of cosmogenesis. Love that moves evolution.

> Christian charity is, by a single stroke and at the same time, dynamized, universalized and (if I may use the term in its most legitimate sense) "pantheized". (*Christianity and Evolution*, 1945, X, 184 E; 215 F)

CHRIST, BODY OF *see also* **Christ, Mystical Body of**
Synonymous with the Mystical Body of Christ.

CHRIST, COSMIC *see also* **Christ, Total; Christ, Universal; Cosmic**
The third or cosmic nature of Christ. Christ Pantocrator. Christ as source and savior of a personal universe where the cosmic is finally assumed by humanity in Christ and transfigured by two inverse (but intimately connected) movements – the ascent of evolution and the descent of grace. Synonymous with the Universal Christ.

(a) Christ has a cosmic body that extends throughout the whole universe. (*Cosmic Life*, 1916, XII, 58 E; 67 F)
(b) Christ coincides with the universe, as the universal center common to cosmic progress and gratuitous sanctification. (*My Universe*, 1918, XIII, 205 E; XII, 304 F)
(c) The Cosmic Christ . . . cannot confine his body to some periphery drawn inside things; he came, above all, for souls and only for souls; he could only bring them together and give them life by assuming and animating, with them, the rest of the world . . . The Gospel of the Cosmic Christ, in which the salvation of the modern world is to be found, is truly the word handed down from heaven to our forefathers. (*Cosmic Life*, 1916, XII, 57–8 E; 67 F)
(d) The great cosmic attributes of Christ, those which (particularly in St. John and St. Paul) accord him a universal and final primacy over creation, these attributes . . . only assume their full dimension in the setting of an evolution . . . that is both spiritual and convergent. (*Catholicism and Science*, 1946, IX, 189 E; 239 F)
(e) I am more and more convinced the Church will only be able to resume its conquering march when (resuming the great theological effort of the first five centuries) it starts to rethink (ultra-think) the relations, no longer between Christ and the Trinity, but between Christ and a universe that has become fantastically immense and organic (at least a thousand billion galaxies each surely containing life and thought . . .). Christianity can only survive (and super-live) . . . by sub-distinguishing in the "human nature" of the Word Incarnate between a "terrestrial nature" and a "cosmic nature". (Letter to Bruno de Solages OCarm, 2 January 1955, *Lettres intimes*, 450)
(f) The modernist "volatilizes" Christ and dissolves him in the world. While I am trying to concentrate the world in Christ. (Cahier 7, 9 June 1919, in Bruno de Solages OCarm, *Teilhard de Chardin*, 342)

(g) Christ is still the only cosmic element capable . . . of embodying modern hopes of a spiritual organization of the world. But here again the Church, recapturing the vision passionately described by St Paul, must be willing to recognize and proclaim him, wholly and entirely, endowed with universal hope and energy. (*The Road of the West*, 1932, XI, 55 E; 60 F)

CHRIST, EUCHARISTIC *see also* **Eucharist**

Christ truly present in the bread and wine of the Eucharist. The Real Presence.

At every moment the Eucharistic Christ controls, from the point of view of the organization of the Pleroma (that is the only true point of view for understanding the world), the whole movement of the universe – the Christ "*per quem omnia, Domine, semper creas, vivificas et præstas nobis*". (*The Divine Milieu*, 1926–7, 1932, IV, 114–15 E; 153 F)

CHRIST, EVER GREATER

(a) I believe I can discern, emerging more solidly than ever, faith in that which is the very axis of christianity: the existence of a personal center (or, rather, an ultra-personal center) of the universe, gradually taking shape ahead of us in the figure of the Ever Greater Christ. I tried last winter to express this vision in a paper *Christology and Evolution* that has been sympathetically received by Fr. Valensin . . . (Letter to Christophe Gaudefroy, 14 July 1934, *Lettres inédites*, 104)

(b) O Ever Greater Christ! (*The Heart of Matter*, 1950, XIII, 58 E; 70 F)

CHRIST, EVOLUTIONARY *see also* **Christ-the-Evolver**

CHRIST, HEART OF *see also* **Golden Glow**

Luminous fringe revealing the divinity of Jesus (epiphany and transfiguration) and the active presence of the divine milieu. The Heart of Jesus at the heart of matter.

(a) The Heart of Christ at the heart of matter. The "Golden Glow" as I like to say in English. (Letter to Jeanne-Marie Mortier, 30 October 1948, *Lettres à Jeanne Mortier*, 48)

(b) The Golden Glow: the glimmer of fire – love (attraction) and evolution of love – the all, centered and non-diffused (expansion). (Journal, 24 September 1948, in Claude Cuénot, *Nouveau Lexique*, 96)

(c) The Heart of Christ universalized coincides with the heart of matter amorized. (*The Heart of Matter*, 1950, XIII, 49 E; 60 F)

CHRIST, HISTORIC

Christ in history/as historical phenomenon within the cosmic order.

The one clear thing to which I should like to devote myself as intensely as possible in my remaining years is to "christify" . . . evolution (that supposes scientific work to determine the "convergence" of the universe and religious work to bring out the universal nature of the Christ of history). (Letter to Marguerite Teillard-Chambon, 8 November 1953, *Letters from a Traveler*, 290; *Lettres de voyage*, 397)

CHRIST, MYSTICAL

(a) The Mystical Christ has not yet attained his full growth. Nor, therefore, has the Cosmic Christ. Both, at the same time, are and are becoming: and the ultimate moving force of all created activity is to be found in the continuation of this engendering . . . Christ is the term of even the natural evolution of living beings; evolution is holy. (*Cosmic Life*, 1916, XII, 59 E; 69 F)

(b) The mystical Christ, the Universal Christ of St Paul has neither meaning nor value in our eyes except as an expression of the Christ who was born of Mary and who died on the Cross. (*The Divine Milieu*, 1926–7, IV, 105 E; 141 F)

CHRIST, MYSTICAL BODY OF

Synonymous with the Body of Christ. Not to be confused with "the initial Body, the primary Body of Christ" (*My Universe*, 1924, IX, 65 E; 93 F) that represents the Body of Christ as God and Man. Animated spiritual organism in a state of perpetual growth. Permits the physical incorporation of the faithful in Christ. Composed of all those who truly participate, through grace, in the divine life of Christ. Organism that will only be completed at the end of time.

(a) Across the enormity of time and the disconcerting multiplicity of individuals, one operation is going on – the incorporation of the elect in Christ; one thing is being made – the Mystical Body of Christ – from all the spiritual powers scattered roughly around the world. (*The Divine Milieu*, 1926–7, 1932, IV, 136 E; 181 F)

(b) An initial justification of the christian attitude is the witness, with St. John and St. Paul, to the "cosmic" wonders of the heavenly Jerusalem: the union of souls in a wonderful organism, the Body of Christ. (Letter to Victor Fontoynont SJ, 15 March 1916, in Henri de Lubac SJ, *Religion of Teilhard*, 244; *Pensée religieuse*, 350)

CHRIST, TOTAL *see also* Christ, Cosmic; Christic

The Christ of St. Paul as consummator of the unity of the universe when "God will be all in all" (1 Cor 15:28). In sum, the Final Christ.

> Between the Word on the one hand and the Man Jesus on the other a sort of christic "third nature" (if I dare say so) emerges – found everywhere in the writings of St. Paul: the total and totalizing Christ in whom, by the transforming effect of the resurrection, the individual human element born of Mary finds itself carried not only to a state of cosmic element (or milieu or curve) but also of final psychic center of universal concentration. (*My Fundamental Vision*, No. 31, 1948, XI, 198–9 E; 214 F)

CHRIST, UNIVERSAL *see also* **Christ, Cosmic; Christ, Total**

Christ as he is, here and now, co-extensive with the universe – even though the consummation of the all will only be achieved in the Total Christ. The Christ of St. Paul and St. John. Synonymous with the Cosmic Christ.

(a) A Christ who only extends to part of the universe, a Christ who did (would) not in some way assume the world would seem to me to be a Christ . . . less grand, less dominant, than the universe of our experience. (*My Universe*, XIII, 1918, 201 E; XII, 300 F)

(b) Christ the King and the Universal Christ. Between the two no more than a simple nuance, perhaps, but an all-important one nevertheless; all the difference between an external power, that can only be juridical and static, and an internal domination that, inchoate in matter and culminating in grace, operates on us by and through all the organic connections of a world in progress. (*The Awaited Word*, 1940, XI, 99 E; 107 F)

(c) The Universal Christ is Christ the organic center of the whole universe . . . on which, in the final analysis, all development, even natural, is physically dependent . . . not only earth and humanity, but also Sirius and Andromeda, the angels and all reality on which we physically depend, near or far (that is, probably all participated being). (*Note on the Universal Christ*, 1920, IX, 14 E; 39 F)

CHRIST-OMEGA *see also* **Omega**

Christ coinciding with both the theological notion of universal center and the notion of the point of ultimate convergence of cosmic evolution. Christ as alpha and omega, plenitude and plenifier, principle and end, who brings about the coincidence between the christic universal center, fixed by theology, and the cosmic universal center, the term of evolution postulated by anthropogenesis. The constitution of the Mystical Body takes place in concordance (sic) with the natural convergence of the world.

> In "Christ-Omega" the universal takes shape and appears in personal form. Biologically and ontologically speaking, there is nothing more coherent and, at the same time, nothing more daring than this identification envisaged at the upper limits of noogenesis between the apparently contradictory properties of the whole and the element. (*My Fundamental Vision*, No. 37, 1948, XI, 203 E; 218 F)

CHRIST-PANTOCRATOR *see also* **Christ, Cosmic; Christ, Total; Christ, Universal**

Christ in glory. Christ as Lord of the universe, Alpha and Omega of history, center of universal concentration . . .

> O Jesus, disperse the clouds with your lightning! Show yourself to us as the Mighty, the Radiant, the Risen! Come to us once again as the Pantocrator who filled the solitude of the cupolas of the ancient basilicas! (*The Divine Milieu*, 1926–7, 1932, IV, 118 E; 158 F)

CHRIST-THE-EVOLVER *see also* **Christ-the-Redeemer; Cosmogenesis**
Christ as supreme mover of cosmogenesis, term and motive force of evolution, attainable in and through the whole process of evolution, that is, by animating the total evolution of the cosmos from within.

(a) Christ-Omega. Christ, animator and collector of all the biological and spiritual energies developed by the universe. Finally, Christ-the-Evolver. (*Super-Humanity, Super-Christ, Super-Charity*, 1943, IX, 167 E; 212 F)
(b) Christ-the-Redeemer . . . is completed, without attenuating in any way his suffering face, in the dynamic plenitude of Christ-the-Evolver. (*Christ the Evolver*, 1942, X, 147 E; 172 F)

CHRIST-THE-HUMANIZER
Christ as promoter of every human value. Christ as he who opens out and completes, through his incarnation, the human person and thereby prepares the human and cosmic nature for the parousia and supernature.

"Plan of work". In 4 parts: a physics . . . a dialectics . . . a metaphysics . . . a mysticism (evolutionary charity, Christ-the-Humanizer, Mysticism of the West). (Journal, 2 September 1947, in Claude Cuénot, *Nouveau Lexique*, 71)

CHRIST-THE-REDEEMER *see also* **Christ-the-Evolver**
Christ as he who, by a positive act of redemption, redresses the human evolutionary current led astray by sin thereby enabling it to participate in the plenitude of the divine life.

(a) Christ-the-Redeemer . . . is completed, without attenuating in any way his suffering face, in the dynamic plenitude of Christ-the-Evolver. (*Christ the Evolver*, 1942, X, 147 E; 172 F)
(b) Christ-the-Redeemer is completing and unfolding himself in the figure of Christ-the-Evolver. (*Introduction to the Christian Life*, 1944, X, 163 E; 191 F)

CHRISTIANITY *see also* **Ecumenism**
Religion of the future.

(a) A general convergence of religions on a Universal Christ who fundamentally satisfies them all: this seems to me the only possible conversion of the world and the only imaginable form for a religion of the future. (*How I Believe*, 1934, X, 130 E; 150 F)
(b) I should be happy to see you . . . substituting for a metaphysics that is stifling us an ultra-physics in which matter and spirit would be englobed in one and the same coherent and homogeneous explanation of the world . . . Thought will explode or evaporate unless the universe, in response to hominization, becomes divinized in some way . . . Christianity is the only living phylum that retains a divine personality. (*Letter to Christophe Gaudefroy*, 11 October 1936, *Lettres inédites*, 110–12)

CHRISTIC *see also* **Centration; Centric; Christ, Total; Union**
One of two evolutionary process: cosmic (natural) and christic (supernatural). Dilation and universalization of the historical Christ to the limits of space and time. In a christic perspective, the historical Christ acquires his total dimension. In an evolutionary (or teilhardian) perspective, the christic implies a fundamental energetic and transforming quality. The christic represents both a property of the universe and the spiritual sense that allows the believer to live by faith in the divine milieu. The cosmic, the human and the christic constitute the three components of teilhardian interior life at the same time as the three dimensions of the real represent the increasingly higher forms of a single movement of centration and union. Teilhardian "keyword".

(a) Between the Word on the one hand and the Man Jesus on the other a sort of christic "third nature" (if I dare say so) emerges – found everywhere in the writings of St. Paul: the total and totalizing Christ in whom, by the transforming effect of the resurrection, the individual human element born of Mary finds itself carried not only to a state of cosmic element (or milieu or curve) but also of final psychic center of universal concentration. (*My Fundamental Vision*, No. 31, 1948, XI, 198–9 E; 214 F)

(b) A remarkable and astonishing region, indeed, where through the meeting of the cosmic, the human and the christic, a new area, the centric, opens up, where the multiple oppositions that go to make up the unhappiness or the anxieties of our existence tend to disappear . . . A miraculous specific effect of the centric that neither dissolves nor subjugates the elements it brings together but personalizes them: precisely because its way of absorbing them is always to "centrify" them more and more. (*The Heart of Matter*, 1950, XIII, 49–50 E; 61–2 F)

(c) It is here the quality of the "christic" bursts into view, such as we saw earlier engendered by the progressive merging together, in our consciousness, of the cosmic demands of an incarnate word and the spiritual potentialities of a convergent universe. (*The Christic*, 1955, XIII, 98–9 E; 113 F)

(d) It seems I am becoming aware of the extraordinary focal point of spiritual radiation concentrated by two thousand years of history in this place: today, it is here [in Rome] that we find the christic pole of the earth; it is here that the ascending axis of hominization is to be found. (Letter to Marguerite Teillard-Chambon, 7 October 1948, *Letters from a Traveler*, 245; *Lettres de voyage*, 338)

CHRISTIFICATION *see also* **Christify**
Process of growing into or becoming absorbed by Christ.

Completely absorbed by the joy of seeing everything around me simultaneously centered, consolidated and amorized, for a long time I only saw, in the immense phenomenon of christification revealed by the coming together of the world and God, the rise within me of the forces of communion. (*The Heart of Matter*, 1950, XIII, 52 E; 64 F)

CHRISTIFY *see also* **Christification**

Action by which God, in the Second Person of the Trinity, becomes human through the incarnation. By the same movement of christification, God christifies the universe, that is, impregnates it with the word.

(a) It is this absence of ontological coherence (and, hence, of spiritual authority) that has just been corrected by the discovery of a universe where . . . God cannot appear as prime mover (ahead) without becoming incarnate and atoning, without becoming christified before our eyes. (*From Cosmos to Cosmogenesis*, 1951, VII, 263 E; 272 F)

(b) With the universe christified (or, that comes to the same thing, with Christ universalized) an evolutionary super-milieu appears – that I have called "The Divine Milieu". (*The Christic*, 1955, XIII, 95 E; 110 F)

CHRISTO-COSMIC *see also* **Christ, Cosmic**

Expression representing the intimate union of Christ and the cosmos.

Today, when I re-read the frankly fervent pages of "The Divine Milieu" I am amazed to find how, already at that time, all the essential characteristics of my christo-cosmic vision were already determined. (*The Heart of Matter*, 1950, XIII, 47–8 E; 59 F)

CHRISTOGENESIS

Constitution of the sphere of Christ. Fourth (and final) phase of evolution where Christ is seen as organic and physical center of a universe in process of becoming in God – in cosmogenesis. Cosmogenesis becomes a christogenesis in a radical perception that organically unites the above of christian faith with the ahead of cosmic evolution. Christogenesis ensures and restores the supernatural destiny of the whole evolutionary process. Teilhardian "keyword".

1 Genesis of the Mystical Body, the Total Christ, the Pleroma. Ultimate expression, through cosmogenesis, of anthropogenesis. Growth of the People of God into the organic unity of the Mystical Body of Christ as understood in Pauline and Johannine theology. The christogenesis of St. John and St. Paul is nothing other than the extension of noogenesis in which cosmogenesis culminates.

2 Essence of christian faith. Christ organically invested with the majesty of his creation. Union with Christ in the Eucharist in a christogenesis means, ipso facto and inevitably, a progressively fuller incorporation in a christogenesis that is none other than the soul of a universal cosmogenesis.

(a) All the communions of our lives are no more than successive instances or episodes of one single communion, of one and the same process of christification . . . The Eucharist, considered in its totality, is none other than the expression and manifestation of the divine unifying energy applied in detail to each spiritual atom of the universe. In short, to adhere to Christ in the Eucharist is inevitably and ipso facto to be incorporated a little more each time in a christogenesis that is none other (and this, as we have seen, is the essence of the christian faith) than the soul of the universal cosmogenesis. (*Introduction to the Christian Life*, 1944, X, 166 E; 194–5 F)

(b) What, when all is said and done, makes christianity immeasurably superior to all other faiths is its increasingly conscious identity with a christogenesis, with the perceptible rise of a certain universal presence, at once immortalizing and unifying. (*The Christic*, 1955, XIII, 90 E; 104 F)

CHRISTOSPHERE

Sphere of Christ. Center of the noosphere. Sphere of the amorizing and transforming omnipresence of Christ. Sphere of spheres. Teilhardian "keyword".

> The next thing I shall write "for myself" (and for my close friends) will perhaps be a study on "the christosphere" or on the christic (christic point, milieu and energies) that more or less brings me back to "The Divine Milieu". (Letter to Jeanne-Marie Mortier, 30 April 1952, *Lettres à Jeanne Mortier*, 97)

CHURCH *see also* **Catholicism; Church, Catholic**

In a phenomenal sense (representing the Church militant), the christic phylum (both linear in time and community of persons in space) by which the supernatural organism, the Mystical Body of Christ, is created and developed.

(a) More explicitly and more truly that before the "crisis" and not simply "by convention" as before, I believe in and love the Church, mediatrix between God and the world. (Letter to Auguste Valensin SJ, 22 August 1925, *Lettres intimes*, 125)

(b) Still deeper, at the very heart of the social phenomenon, we see a sort of ultra-socialization whereby the Church is slowly taking shape, animating by its influence and bringing together all the spiritual energies of the noosphere; the Church, the principal focal point of inter-human relationships through super-charity; the Church, the central axis of universal convergence and precise point of contact between the universe and Point Omega. (*My Fundamental Vision*, No. 24, 1948, XI, 191–2 E; 206 F)

(c) No longer simply the teaching Church but the living Church: germ of super-vitalization planted in the noosphere by the historical appearance of Christ Jesus. Not some parasitic organism, duplicating or deforming the human cone of evolution but an even more interior cone, impregnating, occupying and gradually sustaining the rising mass of the world and converging concentrically towards the summit. (*Outline of a Dialectic of Spirit*, 1946, VII, 149 E; 156 F)

(d) I am more and more convinced the Church will only be able to resume its conquering march when (resuming the great theological effort of the first five centuries) it starts to rethink (ultra-think) the relations, no longer between Christ and the Trinity, but between Christ and a universe that has become fantastically immense and organic (at least a thousand billion galaxies each surely containing life and thought . . .). (Letter to Bruno de Solages OCarm, 2 January 1955, *Lettres intimes*, 450)

(e) The Church is phyletically essential to the completion of the human. (Letter to Pierre Leroy SJ, 15 August 1951, *Lettres familières*, 107; *Letters from My Friend*, 98)

CHURCH, CATHOLIC *see also* Catholicism; Church

Church whose head is the bishop of Rome, successor of St. Peter and St. Paul.

(a) The privilege claimed by the Roman Church of being the sole authentic expression of christianity is far from being an unjustified pretension but is a response to an inevitable organic need. (*Introduction to the Christian Life*, 1944, X, 167 E; 195–6 F)

(b) Without Rome, monotheism would disappear from the noosphere. (Letter to René d'Ouince SJ, 30 March 1950, *Lettres intimes*, 379–80)

(c) Evidence suggests that if christianity is indeed destined, as it professes and feels, to be the religion of tomorrow, it is only through the organized and living axis of its Roman Catholicism that it can hope to match and assimilate the great modern currents of humanitarianism. To be catholic is the only way of being fully and completely christian. (*Introduction to the Christian Life*, 1944, X, 167–8 E; 196–7 F)

(d) The "sin" of Rome . . . is not to believe in a future, in a completion . . . of the human on earth. (Letter to Bruno de Solages OCarm, 17 January 1954, *Lettres intimes*, 434)

CHURCH, INFALLIBILITY OF THE *see also* Assumption, Dogma of the; Cephalization, Law of

(a) To localize, as catholics do, the permanent organ of phyletic infallibility in the councils, or, by an even more advanced concentration of christian consciousness, in the Pope (formulating and expressing, not his own ideas, but the teaching of the Church) fully conforms with the great law of "cephalization" that governs all biological evolution. (*Introduction to the Christian Life*, 1944, X, 153 E; 181 F)

(b) "Infallibility" of the Church: nothing other than the phyletic direction guiding the christian community along the lines of attraction emanating from the divine center. (Note, 27 September 1948, ref. not known)

CIRCLE *see also* **Sphere**
One of the most powerful religious symbols. Primordial image of humanity that points to the single most important aspect of life – its ultimate wholeness.

(a) The circle of presence . . . The circle of consistence . . . The circle of energy . . . The circle of spirit . . . The circle of person. (*The Mystical Milieu*, 1917, XII, 116, 123, 128, 136 E; 157, 164, 169, 178, 188 F)

(b) Our isolation is only partial in relation to the terrestrial organism that is for a time our common matrix. One and the same influence animates and links together everything that thinks. One single circle embraces all spirit and imprisons nothing . . . O wonderful center! O immense sphere! O God! (*The Great Monad*, 1918, XIII, 191 E; XII, 277–8 F)

CO-, COL-, COM-, CON-
Prefix. Expresses the notions of community and convergence both implicit and explicit in teilhardian thought, e.g.

- co-adjustment, co-condition, co-conscience, co-conscientization, co-conscious, co-consciousness, co-culturation, coefficient, co-element, co-engender, co-evolve, co-extended, co-extensive, co-extensivity, co-existence, co-gravitation, coherence, cohesion, co-invention, coordinate, co-organically, co-original, co-thought, co-reflection, co-reflective, coreflectively, co-spiritualization . . .
- collectivization . . .
- communion, complexification, complexity, complexity-consciousness . . .
- concentration, concordism, confluence, conjunction, conjugation, conscience, con-spiration, contact, convergence . . .

CO-CONSCIENTIZATION *see also* **Co-reflection; Conscientization**
Collective progress of consciousness brought about by co-reflection.

> Let us try . . . to put the process of "co-conscientization" in which we take part into a general world perspective. We then see evidence that is quite clear (and strangely liberating) emerging from the facts – that of knowing that beneath the banality and apparent superficiality of the socio-technical development of the earth it is evolution itself, through its orientation towards the improbable, that extends and accelerates beyond our small individual centers in the direction of a complexity-consciousness of planetary dimensions. (*The Christic*, 1955, XIII, 86 E; 100 F)

COHERENCE *see also* **Concordism**
Coherence is truth. Aspect of wholism.

1 Fundamental unity of the universe. Teilhard sees everything in the

universe having a common origin and working towards a common goal.

(a) Greater coherence is the infallible sign of greater truth. (*Turmoil or Genesis?*, 1947, V, 214 E; 275 F)

(b) Truth is nothing other than the total coherence of the universe in relation to every point within itself. (*Sketch of a Personalistic Universe*, 1936, VI, 54 E; 71 F)

2 On the particular level of the relationship between science and religion, represents harmonization through an interior, autonomous and spontaneous movement. Opposite of concordism.

(a) We must not confuse "concordism" and "coherence". We all know how, in the history of ideas, certain puerile, premature efforts at harmonization by mixing the levels and sources of knowledge have only led to unstable, even grotesque arrangements. But such caricatures of harmony should not blind us to the fact that the basic criterion, the specific sign of truth, is to be able to develop, indefinitely, not only without developing internal contradictions but by forming a positively constructed whole whose elements mutually support and complement one another. On a globe, it would be absurd to confuse the meridians at the equator (concordism) but these meridians must, of structural necessity, meet together again at the pole (coherence). (*My Fundamental Vision*, No. 2, 1948, XI, 165 E; 181–2 F)

(b) Two sources of knowledge: science and revelation. The mistake of theologians: to imagine that the two sources are independent, that is, that we can hope to depend on revelation without continually maintaining its coherence (credibility, complementarity) in relation to reason. (Note, 7 November 1951, *Lettres intimes*, 124)

COILING *see also* **Enfolding**

COLLECTIVIZATION *see also* **Socialization**

Phenomenon by which previously dispersed social groups tend to become harmonized to form a whole. Rather vague term lacking the precision of socialization.

(a) If . . . the social unification of the earth is the state towards which evolution is really taking us, this transformation cannot contradict the clearest achievement of evolution in the course of history – the increase of consciousness and individual freedom. Like any other union, the collectivization of the earth, correctly carried out, must super-animate a common soul in us. (*Sketch of a Personalistic Universe*, 1936, VI, 80 E; 100 F)

(b) In the case of the human, collectivization, super-socialization can only mean super-personalization, that is, in the final analysis, sympathy and unanimity (since the forces of love alone possess the property of personalizing by unifying). (*Super-Humanity, Super-Christ, Super-Charity*, 1943, IX, 160 E; 204–5 F)

COMMUNION *see also* **Eucharist; Eucharistization; Mass, The**

(a) All the communions of our lives are no more than successive instances or episodes of one single communion, of one and the same process of christification . . . The Eucharist, considered in its totality, is none other than the expression and manifestation of the divine unifying energy applied in detail to each spiritual atom of the universe. In short, to adhere to Christ in the Eucharist is inevitably and ipso facto to be incorporated a little more each time in a christogenesis that is none other (and this, as we have seen, is the essence of the christian faith) than the soul of the universal cosmogenesis. (*Introduction to the Christian Life*, 1944, X, 166 E; 194–5 F)

(b) All the communions of a lifetime form but one communion. All the communions of all men and women now living form but a single communion. All the communions of all men and women, past, present and future, form but a single communion. (*The Divine Milieu*, 1926–7, 1932, IV, 113 E; 151 F)

(c) There is a communion with God, and a communion with the earth, and a communion with God through the earth . . . In this first basic vision we begin to see how the Kingdom of God and cosmic love may be reconciled: the maternal bosom of the earth is, in some way, the bosom of God. (*Cosmic Life*, 1916, XII, 14, 62 E; 19, 71 F)

COMPLEMENTARITY *see also* **Coherence**

Let us not to repeat, through ideology or sentimentality . . . the error of feminism or democracy at the beginning. Woman is not man: and it is precisely for this reason that man cannot do without woman. A mechanic is not an athlete, or an artist, or a financier . . . These inequalities, that against all the evidence we sometimes try to deny, may appear damaging as long as the elements are regarded statistically in isolation. Observed, however, from the point of view of their essential complementarity, they become acceptable, honorable and even desirable. (*Natural Human Units*, 1939, III, 212 E; 297–8 F)

COMPLEXIFICATION *see also* **Arrangement; Centro-complexity**

Tendency of the real to construct, in favorable conditions, increasingly elaborate (more closely connected and better centered) forms of organization leading to living organisms and the phenomenon of socialization.

(a) Biology can only develop and fit coherently into the universe of science if we are prepared to recognize in life the expression of one of the most significant and most fundamental movements of the world around us . . . the enormous and universal phenomenon . . . of the complexification of matter. (*Man's Place in Nature*, 1949, VIII, 19 E; 27–8 F)

(b) In human beings . . . life . . . has just arrived, through the influence of convergence, at a paroxysm of power causing, simultaneously and progressively, a rise of organization and consciousness in the universe, that is, the interiorization of matter through complexification. (*On Looking at a Cyclotron*, 1953, VII, 356 E; 375 F)

COMPLEXITY *see also* **Aggregate; Centro-complexity; Complexity, Infinity of**

Product of complexification. Antonym of aggregate.

> As I understand it here, complexity is an organized and, consequently, a centered heterogeneity. In this sense, a planet is heterogeneous but not complex. Two different factors or terms are necessary to represent the complexity of a system: one expresses the number and groups of elements contained in the system; the other, much more difficult to illustrate, expresses the number, the variety and the contraction . . . of connections (density) existing between these elements at a minimum volume. (*Man's Place in the Universe*, 1942, III, 222 E; 313 F)

COMPLEXITY, INFINITY OF *see also* **Infinity, Third**

Teilhardian third infinity of complexity that completes the two pascalian infinities – the infinitely large and the infinitely small. Part of the process of complexification and unification that begins with an indefinite multiple and ends in the infinity of the divine center. The infinitely complex is both a third dimension and the synthesis of the first two infinities because its temporal complexifying function is linked to a space that was previously considered static. Synonymous with the third infinity.

> (a) Spreading through a counter-current across entropy, we find a cosmic drift of matter towards increasingly centro-complexified states of arrangement (in the direction of – or interior to – a "third infinity", an infinity of complexity, as real as the infinitesimal and the immense). (*My Phenomenological View*, 1954, XI, 212 E; 233 F)
>
> (b) This "eu-corpusculization" . . . develops "transversally" towards the very small and the very large, in a special form of infinity as real as those (the only ones usually considered) of the infinitesimal and the immense: the third infinity of organized complexity. (*The Singularities of the Human Species*, 1954, II, 212–14 E; 300–2 F)

COMPLEXITY, PRINCIPLE OF THE GREATEST *see also* **Neguentropy**

Evolutionary law expressed through the multiplication and differentiation of relationships between the elements of a whole.

> We envisage as the basis of cosmic physics the existence of a sort of second entropy (or "anti-entropy") bearing, as the effect of chances taken, a part of matter in the direction of increasingly higher forms of structurization and centration. (*Transformation and Continuation in Man of the Mechanism of Evolution*, 1951, VII, 302–3 E; 317 F)

COMPLEXITY-CONSCIOUSNESS *see also* **Complexity-consciousness, Law of**

Correlation of psychic energy (spirit) with a proportionately greater concentration of physical energy (matter). Corresponds to hierarchy of being or doctrine of participation in scholastic thought.

> It is the cosmic axis, of both physical arrangement and psychic interiorization, revealed by the basic drift or orthogenesis, to which I shall constantly refer, every time it is necessary to evaluate the significance of an event or process as an absolute value. Axis of complexity-consciousness, I shall call it, usefully transposable . . . into axis of cephalization (or cerebration) following its appearance in the nature of nervous systems. (*The Singularities of the Human Species*, 1954, II, 215 E; 304 F)

COMPLEXITY-CONSCIOUSNESS, LAW OF *see also* **Complexity, Infinity of; Parameter**

Specifically teilhardian law that expresses the relationship between the laws of thermodynamics and the laws of consciousness. Parameter of the whole of evolution with each stage being evaluated according to two closely correlated factors:

- the degree of complexity, that is, the "organic" inter-linkage between increasingly numerous elements and
- the degree of "consciousness", that is, centric and psychic emergence. Represents a higher and specific form of the process of centration beginning with the emergence of life.

> Life is apparently nothing other than the privileged exaggeration of a fundamental cosmic drift . . . that we can call the "law of complexity-consciousness" and express as follows: "Left long enough to itself, through the extended universal play of chance, matter reveals the property of arranging itself in increasingly complex groups and, at the same time, in increasingly deeper levels of consciousness; this double, conjugated movement of physical involution and psychic interiorization (or centration) is continued, accelerated and pushed as far as possible, once it has begun". (*The Phyletic Structure of the Human Group*, 1951, II, 139 E; 195 F)

COMPRESSION, SOCIALIZATION OF *see also* **Expansion, Socialization of; Noosphere**

Second phase of socialization that emerged in 19th century and consists of a growing mutual pressure of human beings originating in their multiplication and contraction and resulting, through contact, in the formation of an organic and psycho-spiritual whole – the noosphere.

> This evolutionary situation (attached to a still much more generalized move-

ment of "enfolding" cosmic in scope) remained unperceived as long as human socialization remained in its initial phase of expansion (overall occupation of the earth). But it becomes increasingly recognizable as the phase we have now entered – socialization of compression – develops around us. (*The Essence of the Democratic Idea*, 1949, V, 239 E; 310–11 F)

CONCORDISM *see also* **Coherence**

Opposite of coherence. Process that seeks to reconcile dogma and science from outside. Teilhard rejects concordism in favor of the spontaneous and intimate convergence of faith and science (coherence).

(a) Religion and science clearly represent, on the mental plane, two different meridians that it would be wrong not to separate (concordist mistake). But these meridians must necessarily meet at some pole of common vision (coherence): otherwise, everything in our field of thought and knowledge would collapse. (*My Intellectual Position*, 1948, XIII, 144 E; 174 F)

(b) Avoid like the plague any form of "concordism" that seeks to reconcile and justify what is possibly an ephemeral form of dogma and what is possibly also an ephemeral stage of the scientific view . . . On the other hand, try . . . to bring out and develop the basic coherence between what can already be regarded as the definitive axes of science and faith respectively. (Letter, 14 April 1953, in Claude Cuénot, *Teilhard de Chardin*, 395–6 E; 477–8 F)

CONFIDENCE, EXISTENTIAL *see also* **Involution; Reversal**

Passage from pessimism (anguish) to optimism through a process of reversal.

(a) Reversal of fear or existential confidence. (*A Phenomenon of Counter-Evolution in Human Biology*, 1949, VII, 191 E; 197 F)

(b) As though by magic, our fear of matter and people is transformed, inverted, in peace, in confidence, and even in existential love (for someone who has the joy of understanding how a focal point of cosmic attraction, if it is to be personalizing, must itself possess its own super-personality). (*A Phenomenon of Counter-Evolution in Human Biology*, 1949, VII, 194 E; 201 F)

CONSCIENTIZATION *see also* **Conscientize; Consciousness**

Rise of consciousness.

From the most infinitesimal and the most unstable nuclear elements to the most developed living beings, nothing exists and nothing is scientifically conceivable in nature, except as a function of an enormous and uniquely conjugated process of "corpusculization" and "complexification", in the course of which the phases of a gradual and irreversible interiorization ("conscientization") of what we call matter (without knowing what it is) can be distinguished. (*The God of Evolution*, 1953, X, 238 E; 286 F)

CONSCIENTIZE *see also* **Conscientization**
Rise of consciousness.

The absolute towards which we are rising can have no other face than that of the all – an all that is purified, sublimated and "conscientized". (*My Universe*, 1924, IX, 43 E; 71 F)

CONSCIOUSNESS *see also* **Centro-complexity; Reflection; Self-consciousness**
Specific property of arranged states of matter. All forms of psychism from the most extended and most elementary to the most concentrated, where consciousness (on the threshold of human psychism) is expressed by "reflective consciousness" or "reflection". Teilhard uses the term freely while respecting both continuity and discontinuity.

(a) The term "consciousness" is taken in its broadest sense to designate every kind of psyche, from the most rudimentary forms of interior perception to the human phenomenon of reflective thought. (*The Human Phenomenon*, 1938–40, I, 25 n. E; 53 n.1 F)

(b) "Consciousness" presents itself and requires to be treated, not as a particular and subsistent kind of entity, but as an "effect", the "specific effect", of complexity. (*The Human Phenomenon*, 1938–40, I, 222 E; 313 F)

(c) The more complex a being . . . the more it is centered upon itself and therefore the more conscious it becomes. In other words, the higher the degree of complexity in a living creature, the higher the consciousness, and, inversely, the higher the consciousness, the higher the complexity. (*Life and the Planets*, 1946, 111 E; 144 F)

CONSCIOUSNESS, CHRISTIC *see also* **Catholicism; Christic**

(a) What gives christianity its particular effectiveness and tonality is the fundamental idea that the supreme focal point of unity is not only reflected in every element of consciousness that it attracts but, in order to bring about final unification, has had to "materialize" itself as an element of consciousness (the historic, christic "I"). In order to act effectively, the center of centers reflected itself on the world as a center (= Jesus). This concept of Christ, not only as prophet and man exceptionally conscious of God, but also as "divine spark", shocks the "modern" mind at first glance as an outmoded anthropomorphism. But we should note:

(i) that the modern reaction against anthropomorphism has gone much too far, to the point of making us doubt a divine ultra-personality . . .
(ii) and that, psychologically, the astonishing power of mystical development displayed by christianity is indissolubly linked to the idea that Christ is historical. Suppress this key element and christianity becomes but one of many "philosophies": it loses all its force and vitality. (*Some General Views on the Essence of Christianity*, 1939, X, 136 E; 157–8 F)

(b) Christianity, by its very essence, is much more than a fixed system, formulated once and for all, of truths that must be accepted and preserved literally. For all its basis in a nucleus of "revelation", it represents, in fact, a spiritual attitude in process of continual development: development of a christic consciousness that keeps pace with and is required by the growing consciousness of humanity. Biologically, it behaves like a "phylum". By biological necessity, therefore, it must have the structure of a phylum, that is, it must constitute a progressive and coherent system of collectively associated spiritual elements. Clearly, hic et nunc, nothing within christianity but catholicism possesses these characteristics. (*Introduction to the Christian Life*, 1944, X, 167–8 E; 196–7 F)

(c) It is here the quality of the "christic" bursts into view, such as we saw earlier engendered by the progressive merging together, in our consciousness, of the cosmic demands of an incarnate word and the spiritual potentialities of a convergent universe. (*The Christic*, 1955, XIII, 98–9 E; 113 F)

CONSCIOUSNESS, EVOLUTION BY GREATER *see also* **Orthogenesis**

Essential direction of evolution carrying all living creatures towards greater concentration and, consequently, increasingly developed psychism.

> Finally, far below, we find evolution by greater consciousness whereby the mass of living creatures, other than fixed or regressive types, are more or less raised towards greater (individual or collective) organization and greater spontaneity. (*The Spirit of the Earth*, 1931, VI, 27 E; 34 F)

CON-SPIRATION

Psychic factor of sympathy by which each element constitutes with every other element a unanimity of love ("a heart-to-heart") while intensifying its own personal density. Inseparable from unity of thought (or co-reflection) and action. Term expressly borrowed from Édouard Le Roy (1870–1954).

(a) Something that, quite naturally, beyond the only intellectual effects of co-reflection envisaged so far . . . leads us to consider the increasing importance apparently reserved in the future for the noospheric phenomena of sympathy or, in an expression beloved of Édouard Le Roy, "con-spiration". (*The Singularities of the Human Species*, 1954, II, 255 E; 352 F)

(b) A particular and generalized state of consciousness is presaged for our species in the future: a "conspiration" in perspectives and intentions. (*The Grand Option*, 1939, V, 57 E; 78 F)

CONVERGENCE *see also* **Divergence; Emergence; Omega**

1 In a general sense, figure of the evolutionary real whose base and starting point is the indefinitely diluted multiple and whose summit and term is an infinitely concentrated point – Point Omega.

By virtue of the total inter-connection of convergence, neither can an elementary ego become reconciled to the christic center without causing the entire global sphere to become more compressed nor, reciprocally, can the christic center even begin to communicate itself to the least of elements without causing the entire layer of things to become more tightly closed on itself. (*The Christic*, 1955, XIII, 95–6 E; 110 F)

2 In a teilhardian sense, second stage of the dialectics of nature (divergence, convergence, emergence). At each level of being, particularly at the level of *Homo sapiens sapiens*, the new multiplicity produced by divergence tends towards integration through the convergent phenomena of arrangement, union and synthesis that lead to emergence.

Zoologically, the human group is nothing other than a normal bundle of phyla where, following the appearance of a powerful field of attraction, the fundamental divergence of evolutionary shoots finds itself dominated by the forces of convergence. (*The Formation of the Noosphere*, 1947, V, 160 E; 206 F)

CONVERGENCE, MYSTICISM OF *see also* **West, Road of the**

Does the mysticism of convergence risk attaching human beings to things? But we should note that it can lead to ascesis and to attitudes that would satisfy the buddhist: it involves us in going beyond things . . . The virtue of renunciation is a virtue of transcendence and purification. (Lecture, 18 January 1933, notes by Jean Bousquet, in Claude Cuénot, *Nouveau Lexique*, 129–30)

CONVERGENCE, PANTHEISM OF *see also* **Differentiation, Pantheism of; Dissolution, Pantheism of**

Opposite of the static pantheism of dissolution to the extent that the evolutionary convergence of the universe completes the all in a final center and increasingly personalizes the self (that is united in the all). Contains two different elements:

1 pantheism of unification;
2 pantheism of union.

Major classification of pantheism.

Pantheism of convergence . . . Pantheism of differentiation . . . Essentially the effect of love . . . God is all in all (St. Paul) . . . The only pantheism capable of animating experimentally our cosmogenesis (of convergence). (Conversation with F. Lafargue, July 1954, in Claude Cuénot, *Nouveau Lexique*, 147)

CONVICTIONS, TEILHARD'S THREE *see also* **Credo, Teilhard's**
Teilhard's declaration of faith as a loyal member of the Society of Jesus (1948).

> I must insist on the fact that . . . the interior attitude I have just described has the direct effect of linking me ever more inescapably to three convictions which form the marrow of christianity: the unique value of the human as the spearhead of life; the axial position of catholicism in the convergent fascicle of human activities; and finally the essential consummating function assumed by the risen Christ at the centre and summit of creation. (Letter to Jean-Baptiste Janssens SJ, 12 October 1951, *Lettres intimes*, 399–400; *Lettres familières*, 115; *Letters from My Friend*, 105)

CO-REFLECTION *see also* **Con-spiration**
Collective, socialized aspect of human reflection. Process that is more intellectual than con-spiration. Emerges in characteristic form in modern times. Present in inchoate form in any valid individual reflection.

> Humans-as-individuals and humans-as-species (or noosphere) can be shown symbolically by an elliptical figure with two foci – one of material arrangement (brain and technology) and the other of mental study (reflection and co-reflection). (*The Singularities of the Human Species*, 1954, II, 241 E; 335 F)

CO-REFLECTIVE *see also* **Co-reflection**
Cumulative outcome of co-reflection.

> Below the oscillations of human history: the accumulation of a co-reflective. (*The Singularities of the Human Species*, 1954, II, 239 E; 332 F)

CORPUSCULIZATION
Process by which the stuff of the world, while retaining its organic unity, tends to constitute small closed systems, both autonomous and interdependent.

> Two fundamental mechanisms of evolution: corpusculization and ramification. (*The Singularities of the Human Species*, 1954, II, 211 E; 299 F)

COSMIC *see also* **Christic**

1 Universe as a whole constituting, in an evolutionary dimension, the first stage of the organization of the multiple towards emergence of the human.
2 Ability to understand the unity of the world as constituting one of the senses (directions) of spirit – the cosmic sense.

> The cosmic (or the evolutionary). (*The Heart of Matter*, 1950, XIII, 16 E; 23 F)

COSMOGENESIS *see also* **Cosmos**

First phase of evolution. Evolution of the universe. Evolutionary universe conceived as a system animated by an orientated and convergent movement. Dynamic concept. Continuous process of creative union advancing in the general direction of complexity-consciousness. Opposite of static universe.

> In the space of two or three centuries . . . the universe no longer appears to us as an established harmony but has definitely taken on the appearance of a system in movement. No longer an order but a process. No longer a cosmos but a cosmogenesis. (*Reflections on the Scientific Probability and the Religious Consequences of an Ultra-human*, 1951, VII, 272 E; 282 F)

COSMOGENESIS, GOD OF

God as creator of an evolutionary world, expressing the creative act in time as a creative transformation immanent in the universe (even though the creative act, both in principle and in itself, is transcendent). Through the incarnation, Christ superanimates the energy of creative transformation from within.

> Previously, a God of the cosmos, an "efficient" creator, appeared sufficient to fill our hearts and satisfy our minds. From now on (and here there is reason enough to look for the underlying origins of modern religious disquiet), nothing other than a God of cosmogenesis, an "animating" God, can satisfy our capacity for worship. (*From Cosmos to Cosmogenesis*, 1951, VII, 262 E; 270 F)

COSMOGENESIS, HUMANISM OF *see also* **Cosmos, Humanism of; Ultra-human**

Evolutionary neo-humanism defined by belief in the existence of an ultra-human.

> At the moment, following several articles by my good friend, Auguste Valensin, I have been thinking vaguely about something on "Humanism and Humanism". Even Fr. Blanchet, in his excellent article in *Études*, still only appears to see humanism "*à la Grecque*" (Plato, the renaissance) as an aesthetic flowering when we are already caught up in an evolutionary neo-humanism dominated (defined) by the conviction that there is an "ultra-human". A humanism of the cosmos is outmoded, outdated, and is being replaced by a humanism of cosmogenesis – not only humans fully grown (the Greeks) but humans fully evolved (us). (Letter to Jeanne-Marie Mortier, 30 March 1955, *Lettres à Jeanne Mortier*, 179)

COSMOGENESIS, SYSTEM OF *see also* **Cosmos, System of the**

Mode of thought taking as its starting point the evolutionary idea of an organic and convergent universe where every element and every event can only appear connected to the temporal development of the whole and where the stuff of the world folds in on itself around a center – omega.

> In a system of cosmos . . . it was extremely difficult, if not impossible (except by intervention of an accident, itself quasi-inexplicable), to justify in our minds the existence of pain and fault in the world. In a system of cosmogenesis, on the other hand, how much longer must we cry out to make ordinary "public opinion" understand that, intellectually (I do not say emotionally) speaking, not only does the problem of evil become soluble – it no longer arises? . . . Evil, a secondary effect, an inevitable by-product, of the process of a universe in evolution . . . An evil no longer catastrophic but evolutionary. (*From Cosmos to Cosmogenesis*, 1951, VII, 259–60 E; 267–8 F)

COSMOGENESIS, VISION OF *see also* **Cosmogenesis, System of**

> Access to a vision of cosmogenesis in the human vision leads to an enlarged, embellished vision of God. (Lecture, 8 April 1951, notes by Claude Cuénot, in idem, *Teilhard de Chardin*, 261, 447 E; 317, xxiv F)

COSMOS *see also* **Cosmogenesis; Cosmos, God of the; Cosmos, Humanism of the; Cosmos, System of the**

Static universe conceived as a fixed or cyclical system. Opposite of cosmogenesis. The word "cosmos" has many meanings in Teilhard's vocabulary. It can equally mean "universe" or "earth" (or "world" or "planet").

> In the space of two or three centuries . . . the universe no longer appears to us as an established harmony but has definitely taken on the appearance of a system in movement. No longer an order but a process. No longer a cosmos but a cosmogenesis. (*Reflections on the Scientific Probability and the Religious Consequences of an Ultra-human*, 1951, VII, 272 E; 282 F)

COSMOS, GOD OF THE *see also* **Creation, Evolutionary; Fixism; Logicalism**

God as creator of a static or cyclical universe, a universe considered complete from the beginning. God is transcendent to such a universe as a worker is transcendent to his work – even though God is present in the creation he keeps in being and the incarnation links creation to the creator.

> Irresistibly, all around us, by every avenue of experience and thought, the universe is being formed organically and genetically. How, in these conditions, could God the Father of two thousand years ago (still the God of the cosmos) fail to be imperceptibly transfigured, through the very effort of our worship, into a God of cosmogenesis, in some focal point or animating principle of an evolutionary creation in which our individual condition appears much less that of a servant who works than an element that unites? (*The Christian Phenomenon*, 1950, X, 202 E; 236–7 F)

COSMOS, HUMANISM OF THE *see also* **Cosmogenesis, Humanism of**

Humanism conceived as a harmonious growth of the individual person or human race in a static or cyclical universe.

COSMOS, SYSTEM OF THE

Mode of thought taking as its starting point the static idea of a cosmos where time implies neither an organizing nor a convergent operation in the sense of a cosmic structure based on:

1 discontinuously interchangeable elements and
2 external relations capable of expressing the hierarchical structure of an ideal plan but not the whole of its internal organic connections.

Perspective rejected by Teilhard.

> In a system of cosmos . . . a disastrous dualism was inevitably introduced into the structure of the universe. On the one hand, spirit – on the other, matter. *(From Cosmos to Cosmogenesis*, 1951, VII, 258 E; 266 F)

CO-SPIRITUALIZATION *see also* **Research**

Concept of religious research derived from the absolute duty of research.

> We belong to a phylum, where there is a process of co-spiritualization . . . I dream of an age in which there will be among the higher commissions of the Church, not only a Holy Office to whittle down, but a committee to study new ideas. It may come about in two or three generations. The idea of religious discovery has failed to penetrate because of a mistaken idea of revelation. (Last conversation with Claude Cuénot, 7 July 1954, in idem, *Science and Faith in Teilhard de Chardin*, 62)

COUNTER-EVOLUTION

Phenomenon caused by something that slows down, halts or even reverses the progress achieved so far by evolutionary energy.

> Original sin, considered on a cosmic basis . . . tends to be confused with the mechanism of creation where it represents the negative forces of "counter-evolution". (*Christ the Evolver*, 1942, X, 150 E; 175 F)

CREATION *see also* **Union, Creative**

1 The divine act (mystery of personal love partly suggested but not totally penetrated by the human mind) that produces both the most extended multiple (unifiable non-being that offers itself in union) and the unitive power that gradually reduces the multiple by integrating it in increasingly complex syntheses. In this sense, to create is to unite (see also *My Fundamental Vision*, No. 29, 1948, XI, 196 E; 211 F). The teilhardian concept of creation excludes:

- all pre-existence of matter and
- a theomachic theology based on the struggle of the one and the multiple (considered as two equivalent principles).

To be created for the universe is to find ourselves in that "transcendental" relationship with God that makes us secondary, participated, steeped in the divine to the very marrow of our being. We have become (despite repeated assertions that creation is not an act in time) used to linking this condition of "participated" being with the existence of an experimental zero in duration, that is, with a recordable temporal beginning. (*Basis and Foundations of the Idea of Evolution*, 1926, III, 134 E; 188 F)

2 The real as product of the divine act. Creation is a process, not an isolated act in time and space.

(a) Evolution . . . is not "creative" as science once believed but is the expression, in time and space, of our experience of creation. (*Man's Place in the Universe*, 1942, III, 231 E; 323–4 F)

(b) Creation, fall, incarnation, redemption, these great universal events no longer appear as fleeting accidents occurring in time (a childish view that is a constant scandal to our experience and reason); all four become coextensive with the duration and totality of the world: they are in some way the (really distinct but physically linked) aspects of one and the same divine operation. (*Note on Some Possible Historical Representations of Original Sin*, 1922, X, 53 E; 69 F)

(c) Creation, incarnation, redemption . . . the three mysteries really become, in the new christology, no more than three faces of one and the same fundamental process, of a fourth mystery . . . to which we should give a name to distinguish it from the other three: the mystery of the creative union of the world or pleromization. (*Suggestions for a New Theology*, 1945, X, 183 E; 213 F)

(d) Creation is necessarily be expressed through evolution . . . Instantaneous creation (just like the creation of an isolated object) appears to me to be a philosophical "absurdity". (Letter to Bruno de Solages OCarm, 2 April 1935, *Lettres intimes*, 302)

CREATION, EVOLUTIONARY *see also* **Spiral; Union, Creative**

Term first used by Count Bégouën in 1879. For Teilhard, evolution is not creative but creation is evolutionary (evolutive). Evolution is the expression of a creation that is both continuous and multiple. Evolutionary creation is wholly compatible with the doctrine of the original fall.

(a) As far as human origins are concerned, science certainly has a lot to discover and catholics a lot to think about. What we can foresee is that the Church will increasingly recognize the scientific validity of an evolutionary form of creation and science will make more room for the powers of spirit, liberty and, consequently, of "improbability" in the historical evolution of the world. (*What Should We Think of Transformism?*, 1930, III, 157 E; 220 F)

(b) Evolution . . . is not "creative" as science once believed but is the expression, in time and space, of our experience of creation. (*Man's Place in the Universe*, 1942, III, 231 E; 323–4 F)
(c) Evolutionary creation and expectation of a revelation. (*Outline of a Dialectic of Spirit*, 1946, VII, 146, 150 E; 152, 157 F)
(d) Evolutionary creation and statistical origin of evil . . . the creative act is only intelligible as a gradual process of arrangement and unification. (*Reflections on Original Sin*, 1947, X, 193–4 E; 226 F)
(e) How, in these conditions, could God the Father of two thousand years ago (still the God of the cosmos) fail to be imperceptibly transfigured, through the very effort of our worship, into a God of cosmogenesis, in some focal point or animating principle of an evolutionary creation? (*The Christian Phenomenon*, 1950, X, 202 E; 237 F)

CREATIONISM *see also* **Fixism**
Opposite of evolution. Immediate creation by God of independent species, a spontaneous generation of things from the inanimate, something scientifically impossible since all living organisms, except at the very beginning, can only be born of the living. Perspective rejected by Teilhard.

CREDO, TEILHARD'S *see also* **Convictions, Teilhard's Three**
Teilhard's own credal statement, first formulated in 1934, strengthened and reaffirmed in 1950.

(a) I believe the universe is an evolution:
I believe evolution proceeds towards spirit;
I believe spirit, in human beings, completes itself in the personal;
I believe the supreme personal is the Universal Christ. (*How I Believe*, 1934, X, 96 E; 117 F; *The Heart of Matter*, 1950, XIII, 78 n.7 E; 38 n.2 F)
(b) I believe in the Church, "mediatrix" between God and the world . . . (Letter to Auguste Valensin SJ, 22 August 1925, *Lettres intimes*, 125)

CROSS
One of four fundamental symbols, with the center, circle and square. Dynamic symbol for Teilhard who speaks of "the prodigious spiritual energetics born of the Cross" (*The Spiritual Energy of Suffering*, 1950, VII, 248 E; 256 F) and sees his personal quest in life as an "impassioned embrace . . . in the arms of the Cross" (*Cosmic Life*, 1916, XII, 68 E; 78 F).

(a) In sum, Jesus on the Cross is together symbol and reality of that immense labor over the centuries that has, little by little, raised up the created spirit in order to concentrate it in the depths of the divine milieu. He represents (and, in a real sense, is) creation that, sustained by God, reascends the slopes of being, sometimes clinging to things for support, sometimes tearing itself from them in order to go beyond them, and always compensating, by physical suffering, for its moral failures. The Cross is, therefore, not something inhuman but super-human . . . Christians are asked to live, not in the shadow of the Cross, but in the fire of its creative action. (*The Divine Milieu*, 1926–7, 1932, IV, 87–8 E; 118–19 F)

(b) Everything changes in an impressive way on the screen of the evolutionary world we have just described. Projected onto such a universe in which the struggle against evil is the *sine qua non* of existence, the Cross takes on new gravity and beauty we find most appealing. No doubt Jesus is always he who bears the sins of the world; moral evil is mysteriously made good by suffering. But, more than this, it is he who surmounts structurally in himself, and for us all, the resistance to spiritual ascent inherent in matter. It is he who bears the structurally inevitable burden of every species of creation. He is the symbol and act of progress. The complete and definitive meaning of redemption is no longer simply to expiate but also to traverse and conquer. The full mystery of baptism is no longer to cleanse but (as the Greek Fathers understood so well) to plunge into the fire of the purifying struggle "for being". No longer the shadow, but the passion of the Cross. (*Christology and Evolution*, 1933, X, 85 E; 103–4 F)

(c) In a universe of cosmogenesis where evil is no longer "catastrophic" (that is, born of an accident) but "evolutive" (that is, a statistically inevitable by-product of a universe in the course of unification in God) – in such a universe the Cross (without losing its expiatory and compensatory function) becomes even more the symbol and the expression of the whole of "evolution" ("noogenesis"): coreflection and unanimisation of humanity through and by means of suffering, sin and death . . . Christ on the Cross is the most complete expression of a "God of Evolution" in human consciousness. A God of Evolution, that is, a divinizing, christifying God, both above and ahead. (Letter to André Ravier SJ, 8 April 1955, *Lettres intimes*, 465)

CROSS, DOCTRINE OF THE

(a) The doctrine of the Cross is that to which all people adhere who believe that the immense human agitation with which they are faced opens on to a road that leads somewhere and that this road rises. Life has a term: therefore it imposes a direction, a direction oriented, in fact, towards the highest spiritualization through the greatest effort . . . (*The Divine Milieu*, 1926–7, 1932, IV, 85 E; 116 F)

(b) On the Cross, we are perhaps in danger of seeing only an individual suffering and a simple expiation. The creative power of which death escapes us. Let us take a broader view and we shall see that the cross is the symbol and focal point of an action of inexpressible intensity. Even from the earthly point of view, fully understood, the crucified Jesus is neither rejected nor conquered. It is, on the contrary, he who bears the weight and draws ever higher towards God the universal march of progress. (*The Significance and Positive Value of Suffering*, 1933, VI, 51–2 E; 65–6 F)

CROSS, SENSE OF THE

(a) The Cross has always been a sign of contradiction and a principle of selection among people . . . Only too often the Cross is presented for adoration, not so much as a sublime end to be attained by our transcending ourselves, but as a symbol of sadness, restriction and repression. (*The Divine Milieu*, 1926–7, 1932, IV, 84, 85 E; 115 F)

(b) This idea of a value of sacrifice and pain for the sake of sacrific and pain itself . . . is a dangerous (and very "protestant") perversion of the "meaning of the Cross" (the true meaning of the Cross is: "Toward progress through effort") . . . (Letter to Rhoda De Terra, 18 September 1948, Or. Eng., *Letters to Two Friends*, 187)

(c) Christians are asked to live not in the shadow of the Cross but in the fire of its creative action. (*The Divine Milieu*, 1926–7, 1932, IV, 88 E; 119 F)

(d) The full mystery of baptism is no longer to cleanse but (as the Greek Fathers saw only too well) to plunge into the fire of the purifying struggle "for being". No longer the shadow, but the passion of the Cross. (*Christology and Evolution*, 1933, X, 85 E; 104 F)

CULTURAL, THE

Biological differentiation of species, starting at the human level, by reflective differentiation of modes of thought and civilization. In this sense, the spiritual reality of a culture is represented by the continuation of the specific in reflective form and the emergence of the new threshold – the threshold of the personal. One can speak of a "tree" of civilization as one speaks of a tree of life.

> Not only . . . are humans a species like other species but also, and above all, more of a species than other species because:
> (a) firstly, they represent a species that has biologically emerged (into the reflective);
> (b) secondly, in them, following this emergence, speciation works on a new level (the "cultural") . . . (*Hominization and Speciation*, 1952, III, 260 E; 370 F)

CURVE *see also* **Evolution**

> Drawn all along together by a single fundamental current, one after the other all the domains of human knowledge have set off toward the study of some kind of development. Does this mean evolution is a theory, a system, or a hypothesis? Not at all: yet something far more. Evolution is a general condition, that all theories, all hypotheses, all systems must submit to and satisfy from now on in order to be conceivable and true. Evolution is a light illuminating all facts, a curve that every line must follow. (*The Human Phenomenon*, 1938–40, I, 152 E; 242 F)

CYBERNETICS *see also* **Reflection**

Theoretical (and interdisciplinary) study of control and communication in machines, industry, society and ecology.

> Nothing characterizes modern times more surely and more exactly than the irresistible invasion of the human world by technology. Mechanization, invading, like a tide, every part of the world and every form of social activity. Mechanization, rapidly overflowing the limits of individual, provincial and national tasks to rise to the level of planetary operations. Mechanization, only recently passing the stage of taking over and multiplying mechanical efforts to assume the same function in the mental domain. All the beginnings of cybernetics with its prodigious possibilities of automatic combination and communication. (*The Phyletic Structure of the Human Group*, 1951, II, 160 E; 221 F)

CYCLE *see also* **Circle; Spiral**

Symbol of a static universe. The cyclical in a static universe becomes a spiral in an evolutionary universe.

> Within the field of our experience, the concrete material universe seems unable to continue on its course indefinitely. Instead of moving indefinitely in a closed cycle, it irreversibly describes a branch with a limited development. And through this it separates itself from abstract magnitudes, to take its place itself among the realities that are born, that grow and that die. (*The Human Phenomenon*, 1938–40, I, 20 E; 47 F)

D

DANTE

Dante Alighieri (1265–1321), poet and mystic. Teilhard's Beatrix in "The Eternal Feminine" (1918) is inspired by Dante's Beatrice. He shares with Dante a strong visual imagination. The image of the circle is particularly persistent in both. "There can be little doubt that if Dante were writing the "Paradiso" today, Teilhard de Chardin would shine forth among that double circle of lights that are the souls of those who sought to reconcile the truth of man with the truth of God . . . It is remarkable that a poet and a scientist, separated by over 600 years and approaching the subject from what would seem to be totally opposed points of view, should both use the sphere as the image of the universe and the "Point Beyond" as the image of God or Omega" (Barbara Reynolds, intro., Dante, *The Divine Comedy, Paradise*, 1962, 30).

> I see . . . there is a translation of Dante's *Vita Nuova* by Henry Cochin. This reminds me that one of the most interesting mystics to study from my point of view would, in fact, be Dante who is so impassioned and so captivated by the real. I believe in any case there are few better examples than Beatrice to make one understand what is meant by the aggrandizement (to the dimensions of the universe) of the sentiment nourished by a particular object (and of the object itself). (Letter to Marguerite Teillard-Chambon, 10 June 1917, *The Making of a Mind*, 195; *Genèse d'une pensée*, 254)

DARWINISM

Teilhard sees darwinian evolution as mechanistic and unguided. If darwinism is correct, then pessimism about the future of the human species in the evolutionary process is justified and Sartre is right in seeing life as utterly absurd and meaningless. Teilhard, however, believes evolution has meaning. It is guided. It is finalistic and eschatological (Francis Klauder SDB, *Aspects of the Thought of Teilhard de Chardin*, 59).

> We should say that, beginning with the human species, biological evolution, not only rebounds (on a new scale and with new resources), but also rebounds reflectively on itself. The darwinian era of survival by natural selection (the vital thrust) is being succeeded by a lamarckian era of superlife by calculated invention (the vital impulse). In the human species evolution is interiorized and finalized . . . (*The Human Rebound of Evolution*, V, 1947, 212 E; 271 F)

DEATH

Awareness of death is a fundamental fact of human existence. Death is "a complete organic transformation" (Letter to Marguerite Teillard-Chambon, 12 July 1918, *The Making of a Mind*, 214; *Genèse d'une pensée*, 278).

(a) Death surrenders us totally to God; it makes us enter into him; we must, in return, surrender ourselves to death with absolute love and abandonment – since, when death comes, the only thing we can do is to let ourselves be completely dominated and guided by God. (Letter to Marguerite Teillard-Chambon, 13 November 1916, *The Making of a Mind*, 145; *Genèse d'une pensée*, 186)

(b) The function of death is to provide the necessary entrance into our inmost selves. It will make us undergo the required dissociation. It will put us into the state organically needed if the divine fire is to descend upon us. And in that way its fatal power to decompose and dissolve will be harnessed to the most sublime operations of life. What was by nature empty and void, a return to plurality, can become in every human existence fullness and unity in God. (*The Divine Milieu*, 1926–7, 1932, IV, 68–9 E; 94 F)

DECENTRATION *see also* **Centration; Super-centration**

Second stage of the existential dialectic of happiness. Openness to the other. Ascetic action that transcends individual centration without denying the achievement and facilitates the perception, first, of the intimate links between personal centers and, second, of dependence on the divine center.

(a) If we are to be fully ourselves and fully alive, we must:

(1) be centered on ourselves;
(2) be decentered on the "other";
(3) be super-centered on one greater than ourselves . . .

Not only physically, but intellectually and morally, we are only human if we develop ourselves. (*Reflections on Happiness*, 1943, XI, 117 E; 129–30 F)

(b) If we are to be happy . . . we must react against the egoism that pushes us either to close up on ourselves or to dominate others . . . The only truly beatifying love is expressed through commonly achieved spiritual progress. (*Reflections on Happiness*, 1943, XI, 125 E; 136–7 F)

DESCENT *see also* **Ascent**

Moral dimension. Antonym of ascent.

Properly speaking, there are neither sacred nor profane things, neither pure nor impure things. There is only a good direction and a bad direction: the direction of ascent, of expanding unification, of greater spiritual effort; and the direction of descent, of constricting egoism, of materializing enjoyment. (*The Evolution of Chastity*, 1934, XI, 72 E; 78 F)

DETACHMENT *see* **Traversing, Detachment by**

DETACHMENT, MYSTICISM OF *see* **West, Road of the**

DÉTENTE, PANTHEISM OF *see* **Diffusion, Pantheism of**

DIALECTICS *see also* **Apologetics; Ethics; Union, Dialectics of**
Reflective method.

1 Deductive philosophical investigation (investigation of the ultimate nature of existence, of knowledge, of good) comprising:
- ontology (nature of existence or being);
- epistemology (nature and validity of knowledge);
- ethics (distinction between "good" and "bad" in human conduct).

2 Deductive metaphysical investigation (investigation of the ontological or ultimate nature of things). Second of four-stage teilhardian thought:
- first: physics (human phenomenon);
- second: dialectics (teilhardian dialectics of nature = divergence, convergence, emergence).
- third: metaphysics (creative union = evolutionary creation, incarnation, redemption);
- fourth: mysticism (evolutionary charity, western mysticism).

(a) Re-write "My Universe". (1) Physics . . . (2) Dialectics: Point Omega, revelation, Christ Omega . . . (3) Metaphysics . . . (4) Mysticism . . . (Journal, 27 August 1947, in Claude Cuénot, *Nouveau Lexique*, 71)

(b) "Plan of work". In 4 parts: a physics . . . a dialectics . . . a metaphysics . . . a mysticism (evolutionary charity, Christ-the-humanizer, mysticism of the west). (Journal, 2 September 1947, in Claude Cuénot, *Nouveau Lexique*, 71)

(c) In the course of efforts to clarify and make the universe around us coherent, the human mind not only works through repeated gropings. It moves forward by successive "comings-and-goings" between the best known and the least known: every move made towards the above, in penetrating the least known, allows it to perceive better (through redescent) the best known in order to rebound, through successive reflections, towards a better understanding of the least known. The mechanism of our vision is made up of oscillating sparks, not a continuous burst . . . In order to avoid any equivocation in the future I think it would be useful to put forward, clearly disarticulated, the successive stages of my apologetics – or, if you prefer, my dialectics. (*Outline of a Dialectic of Spirit*, 1946, VII, 143 E; 149 F)

DIAPHANY, CHRISTIC

Transparency of the universe revealing the presence of Christ to the purified and discerning mind. Fundamental intuition of teilhardian philosophy. Teilhard is wholly anti-manichæan.

> If we might slightly alter a sacred word, we would say the great mystery of christianity is not exactly the appearance but the transparence of God in the universe. Oh yes, Lord, not only the ray that brushes the surface but the ray that penetrates. Not your epiphany, Jesus, but your diaphany. (*The Divine Milieu*, 1926–7, 1932, IV, 121 E; 162 F)

DIFFERENTIATION *see also* **Ultra-differentiation; Union**

Differentiation is what Teilhard calls personalization and Jung calls individuation (letter to Christophe Gaudefroy, 11 October 1936, *Lettres inédites*, 112).

(a) True union (that is, spiritual union or union of synthesis) differentiates the elements it brings together. (*How I Believe*, 1934, X, 117 n.5 E; 137 n.1 F)

(b) The differentiation of beings (the immediate term of their individual perfection) is nothing other than . . . the preparation for an ever closer and more spiritual union of the elements of the universe. (*My Universe*, 1918, XIII, 205 E; XII, 304 F)

(c) True union does not confuse the beings it brings together: it differentiates them more fully; that is, it ultra-personalizes reflective particles. The whole is not the antipodes but the very pole of the person. Totalization and personalization are two expressions of one single movement. (*The Salvation of Mankind*, 1936, IX, 137 E; 178 F)

(d) True union towards the above, in the spirit, ends by establishing in the their own perfection the elements it dominates. Union differentiates. (*Human Energy*, 1937, VI, 144 E; 179 F)

DIFFERENTIATION, EVOLUTION BY INSTRUMENTAL *see also* **Forms, Orthogenesis of**

Dispersive evolution of living creatures characterized by the acquisition of organs that play the part of specialized instruments at the origin of the differentiation of phyla.

> Further down, we find evolution by instrumental differentiation whereby forms are distributed in various "radiations", each one being defined by the acquisition of a specialized morphological type (swimming, running, flying – burrowing creatures, carnivorous animals): most of the "phyla" identified by paleontology are derived from these transformations. (*The Spirit of the Earth*, 1931, VI, 27 E; 33–4 F)

DIFFERENTIATION, PAIN OF

Positive effort and suffering at the basis of the passage to a higher state of concentration or greater complexity.

Unification is a labor. In a very true sense . . . the only support for plurality is the unity ahead. And yet this return to equilibrium is a laborious ascent that only takes place by overcoming true ontological inertia. (*Sketch of a Personalistic Universe*, 1936, VI, 86 E; 107 F)

DIFFERENTIATION, PANTHEISM OF *see also* **Union, Pantheism of**
Synonym of pantheism of union.

God finally becoming all in all within an atmosphere of pure charity ("*sola caritas*"); the very essence of the message of Jesus is unequivocally expressed in this magnificent definition of the pantheism of differentiation. (*Reflections on Two Converse Forms of Spirit*, 1950, VII, 225 E; 234 F)

DIFFUSION, PANTHEISM OF
Synonyms: Détente, Pantheism of; Dissolution, Pantheism of; Effusion and dissolution, Pantheism of; Identification, Pantheism of; Pantheism, Ancient; Pantheism, False; Pantheism, Idealistic; Pantheism, Materialist; Pantheism, Naturalist; Pantheism, Old; Pantheism, Pagan; Pantheism, Popular; Pseudo-pantheism. Major classification of pantheism. Non-evolutionary perspective rejected by Teilhard in which God is the impersonal all in so far as union of the self with God represents a loss of consciousness and a dissolution of the person in the elementary.

Pantheism of diffusion . . . Pantheism of identification . . . Attitude without love . . . God is all . . . This pantheism has no positive evolutionary (or anthropogenetic) component. (Conversation with F. Lafargue, July 1954, in Claude Cuénot, *Nouveau Lexique*, 146)

DIMINUTION, PASSIVITIES OF
Forces that destroy human energies but that can also be the occasion of greater openness to divine action.

The forces of diminution are our real passivities. Their number is vast, their forms infinitely varied, their influence constant. In order to clarify our ideas and direct our meditation we shall divide them into two groups corresponding to the two forms under that we considered the forms of growth: diminutions of internal origin and diminutions of external origin. (*The Divine Milieu*, 1926–7, 1932, IV, 59 E; 81–2 F)

DIRECTION, NEO-ANTHROPOCENTRISM OF *see also* **Movement, Neo-anthropocentrism of**

Since Galileo it might seem that we had lost our privileged position in the universe. Under the increasing influence of the combined forces of invention and socialization, we are now in process of regaining our leadership: no longer in stability but in movement; no longer the center but the leading shoot of a world in growth. Neo-anthropocentrism, no longer of position, but of direction in evolution. (*Evolution of the Idea of Evolution*, 1950, III, 246–7 E; 349 F)

DISPERSION *see also* **Divergence**

The exaggeration of the apparent dispersion of the phyla. (*The Human Phenomenon*, 1938–40, I, 73 E; 127 F)

DISPERSION, EVOLUTION BY

Divergent movement of evolution of living forms that, at the same stage of development, seem to show only unimportant or non-functional variations.

Of the three (types of biological evolutions) the most superficial (one could call it evolution by dispersion) effectively consists of a simple diversification (or spreading) of living forms within a spindle of equal possibilities of shape and color, e.g. certain groups of plants, lepidoptera (butterflies, moths), fish or antelopes. (*The Spirit of the Earth*, 1931, VI, 27 E; 33 F)

DISPERSION, MUTATION BY *see also* **Canalization, Mutation by**

Differentiation of living forms that fails to produce orthogenesis but results in their dispersal within the same genus or family.

By dispersion, firstly, when particular living creatures differentiate in every direction within their type, like a ray of white light that is refracted in a continuum of vibrations on one plane. (*The Movements of Life*, 1928, III, 145 E; 204 F)

DISSOLUTION, PANTHEISM OF

Synonym of pantheism of diffusion

Our generation, essentially pantheist because evolutionist, only seems to understand pantheism in the form of a dissolution of individuals in a diffuse vastness. This is an illusion, due to the fact that the unity of the world is wrongly sought, under the influence of the physical sciences, in the direction of the increasingly simple energies into which it decomposes. (*Sketch of a Personalistic Universe*, 1936, VI, 67 E; 85 F)

DIVERGENCE *see also* **Convergence; Emergence**

First stage of the teilhardian dialectics of nature (divergence, convergence, emergence). At each level of being there is a tendency to dispersion, to the creation of a new multiplicity, that is, of secondary matter that later undergoes a process of convergence or unification.

And from then on, the same two series of effects emerge into the foreground that we called attention to above as we described in its main outlines the behavior of hominization: (1) first of all, the appearance of political and cultural units on top of the genealogical verticils; a complex gamut of group-

> ings . . . (2) and simultaneously the manifestation among these new kinds of branches of the forces of coalescence (anastomoses and confluences) freed in each one of them by the individualization of a psychological sheath – or, more precisely, a psychological axis,. In short, an entire conjugated interplay of divergences and convergences. (*The Human Phenomenon*, 1938–40, I, 143 E; 230–1 F)

DIVINE MILIEU *see also* **Milieu, Divine**

DOGMA

(a) One can only be a christian by believing absolutely and definitively in all the dogmas. (Cahier 1, 1 May 1916, in *Journal*, 73)

(b) The essential quality of the orthodox evolution of dogma is to transform (objectively) divine reality without losing its quality of being objectively given. (Cahier 9, 5 July 1922, in Bruno de Solages OCarm, *Teilhard de Chardin*, 350–1)

DRIFT *see also* **Ascent; Descent**

Fundamental evolutionary process governed by axial energies. Two major drifts can be identified:

- descent towards the probable through the effect of large numbers;
- ascent towards the improbable (vitalization) through the preferential utilization of opportunities offered by the play of large numbers.

The two drifts appear contradictory but are complementary with the appearance of closed systems involving an increase in entropy.

> The more we study this situation, the more we are convinced, in the case of the terrestrial noosphere as in the case of the atoms, the stars or the continents, of certain basic drifts (the very nucleus of the phenomenon) being hidden beneath the veil of cyclical movements especially studied by science up to now – drifts incapable of not always progressing in the same direction, always progressing further, that is, of not producing some specific event of explosion, maturation or transformation. (*The Phyletic Structure of the Human Group*, 1951, II, 168 E; 231 F)

DURATION *see also* **Space–time; Time**

> What makes and classifies someone as "modern" (and scores of our contemporaries are still not modern in this sense) is to have become capable of seeing not only in space or in time, but in duration – or, what amounts to the same thing, of seeing in biological space–time. (*The Human Phenomenon*, 1938–40, I, 152 E; 242–3 F)

DYAD

Platonic term borrowed by Teilhard to represent the spiritualized couple.

For those who know how to see an essential complement to this great cosmic event of reflection in the discovery of the shape of what we might call the "breakthrough to amorization". Even after the flash of individual self-revelation, primitive humans would have remained incomplete if, on meeting the opposite sex, they had not burst into flame in a centric person-to-person attraction completing the appearance of a reflective monad and the formation of an affective monad. (*The Heart of Matter*, 1950, XIII, 60 E; 73–4 F)

EARTH

The word "earth" has many meanings in Teilhard's vocabulary. It can equally mean "world" (or "planet") or "universe" (or "cosmos").

EARTH, SENSE OF THE

Intuition of interdependency (solidarity) with the structure of the earth that leads to humanity forming a real sphere of mind and spirit – the noosphere – around the globe.

> By "sense of the earth" we should understand here the passionate sense of the common destiny that leads the thinking portion of life ever further forward. (*The Spirit of the Earth*, 1931, VI, 31 E; 39 F)

EARTH, SPIRIT OF THE

Human unanimity within which each fully differentiated person will be the (partial but irreplaceable) expression of a spiritual totality that will be specific to the earth.

> How can we determine, by an initial approximation, the higher term to come towards which the transformation, in which we are involved with the world, is leading us? No way . . . other than as a state of unanimity in which each grain of thought, carried to the limits of its particular consciousness, will only be the partial, elementary, incommunicable expression of a total consciousness that is common and specific to the whole earth: a spirit of the earth. (*The Atomism of Spirit*, 1941, VII, 39–40 E; 46 F)

EAST, ROAD OF THE *see also* Identification, Pantheism of; West, Road of the

Eastern mysticism that tends to present spirit as the opponent of matter with unity being obtained, not by a personalizing convergence (unity of tension), but by a suppression of the multiple and dissolution of the person in an impersonal all (unity of détente).

> (a) According to the first way (that I shall call more or less for convenience the "Road of the East"), spiritual unification is conceived as being effected by a return to a common "divine" basis subjacent to, and more real than, all sensibly perceived determinants of the universe. From this point of view, mystical unity appears and is obtained by the direct suppression of the multiple, that is, by relaxation of the cosmic effort of differentiation in and

around ourselves. Pantheism of identification. Spirit of "détente". Unification with the sphere by co-extension of dissolution. (*My Fundamental Vision*, No. 33, 1948, XI, 200 E; 215 F)

(b) The history of western mysticism could be described as a long attempt by christianity to recognize and separate, deep within itself, the eastern and western paths of spirituality: suppress or sublimate? Divinize by sublimation: it was on this side that the profound logic, the instinct of the nascent world, was to go. Divinize by suppression: it was in this direction that the accustomed ways of the ancient east were to push. (*The Road of the West*, 1932, XI, 51–2 E; 57 F)

ECOLOGY

Expresses relationship between living organisms and their environment.

The multiple factors (ecological, physiological, psychic . . .) working to bring together and link firmly living beings in general (and human beings in particular) are only the extension and expression of the forces of complexity-consciousness ever at work, as we said, to build (as far as possible and wherever possible in the universe), contrary to entropy, corpuscular combinations of an ever higher order. (*The Phyletic Structure of the Human Group*, II, 1951, 156 E; 215–16 F)

ECUMENISM *see also* **Christianity**

Teilhard is opposed to any form of syncretism. (Cf. Henri de Lubac SJ, "Teilhard and the Problems of Today," in *The Eternal Feminine*, 178–89).

(a) A general convergence of religions on a Universal Christ who fundamentally satisfies them all: this seems to me the only possible conversion of the world and the only imaginable form for a religion of the future. (*How I Believe*, 1934, X, 130 E; 150 F)

(b) Ecumenism, neither of diffusion nor regression, but of progression in a convergent milieu. (Letter to Jeanne-Marie Mortier, 22 September 1954, *Lettres à Jeanne Mortier,* 163)

EFFUSION AND DISSOLUTION, PANTHEISM OF

Synonym of pantheism of diffusion.

In fact, without realizing it, I had then arrived, in the course of my awakening to the cosmic life, at a dead end from which I could not escape without the intervention of some new force or light. A dead end. Or rather a subtle tendency to drift towards a lower form (banal and facile) of pantheist spirit: pantheism of effusion and dissolution . . . To be all, I had to be fused with all. (*The Heart of Matter*, 1950, XIII, 23–4 E; 31–2 F)

EGO

1 In a general sense, organization of the conscious mind composed of conscious perceptions, memories, thoughts, feelings.

2 In a teilhardian sense, "ego" (in contrast to the "true self") denotes the proud, defiant self-reliance, the attempted autonomy of the human in revolt against God. The true self can only be liberated through the death of the ego. We are only truly ourselves when we replace egocentricity by theocentricity and finds our true selves in the God in whom "we live and move and have our being" (see also Gerald Vann OP, *Hymn of the Universe*, 32–3 n).

> Lord, bury me in the very depths of your heart. And then, holding me there, burn me, purify me, inflame me, sublimate me, until I become wholly what you would have me be, through the utter annihilation of my ego. (*The Mass on the World*, 1923, XIII, 130 E; 152 F)

EGO, NUCLEAR *see also* **Ego, Peripheral**
Psychic tendency to become self-centered and individualized, personal and incommunicable.

> It seems to me the only way out of this difficulty is to imagine two sorts of ego in each phyletic center; a nuclear ego (more or less complete or rudimentary, according to the individual case) and a peripheral ego, incompletely individualized and, consequently, capable of being divided – capable, after separation, of developing new shoots and of isolating within itself a new nucleus of incommunicable ego. (*Centrology*, 1944, VII, 108 E; 114 F)

EGO, PERIPHERAL *see also* **Ego, Nuclear**
Psychic link with the germ – the part of the organism that can be transmitted and can be divided into a series of mother cells that cause the appearance of new nuclear egos.

ELEMENT, UNIVERSAL
Milieu (composed of elements like water and air) initially seen as cosmic, then gradually transfigured by the increasingly felt effect of the christic presence. Ultimately becomes Christ himself.

> Many people . . . feel the need and capacity for apprehending a universal physical element in the world that establishes always and everywhere a relationship between themselves and the absolute – both in and around them. (*The Universal Element*, 1919, XII, 290 E; 431 F)

EMERGENCE *see also* **Convergence; Divergence; Emergence, Principle of; Emersion; Threshold**
Third stage of the teilhardian dialectics of nature (divergence, convergence, emergence). Appearance through synthesis of a wholly new phenomenon that is the product of preceding stages and contingent, that is, contains unforeseen properties and a new specificity.

Is not the very act by which the fine point of our mind penetrates the absolute a phenomenon of emergence? (*The Human Phenomenon*, 1938–40, I, 153 E; 244 F)

EMERGENCE, PRINCIPLE OF
Evolutionary law that links every new form of the real with the antecedents on which it depends.

According to this state of mind, the idea that some soul of souls is in preparation at the summit of the world is not so foreign to current views of human reason as one might think. After all, how else could our thought generalize the principle of emergence? . . . There is only one way for our mind to integrate these two essential properties of the autonomous center of all centers into the coherent pattern of noogenesis and that is to take up the principle of emergence and to complete it. Our experience males it perfectly clear that, during the course of evolution, emergence only happens successively and in mechanical dependence on what precedes it . . . And this is even how omega itself is discovered by us at the end of the process, insofar as the movement of universal synthesis culminates in it. (*The Human Phenomenon*, 1938–40, I, 191–3 E; 299–301 F)

EMERSION *see also* **Emergence**
Appearance through synthesis of a wholly new phenomenon that is both the product of preceding stages and contingent, that is, contains unforeseen properties and a new specificity. Concept similar to emergence but distinguished by two essential characteristics:

1 movement towards the transcendent;

The end of the world: the reversal of equilibrium, detaching the spirit, complete at last, from its material matrix, to rest from now on with its whole weight on God-Omega . . . The end of the world: critical point, simultaneously, of emergence and emersion, of maturation and escape . . . (*The Human Phenomenon*, 1938–40, I, 206 E; 320 F)

2 instantaneous element ("exact" or "perfect") of a critical threshold, as opposed to the enduring element, of a process of emergence.

At the origins of life it seems to be the focal point of arrangement in each individual (F1) that generates and controls its conjugated focal point of consciousness (F2). But we see that higher up the equilibrium is reversed. From the "individual step of reflection" (if not already before), it is very clearly F2 that begins to take charge of the progress of F1 (by "invention"). And then, higher still, that is to say at the (conjectured) approach of collective reflection, we see F2 acting as though it is dissociating from its temporo-spatial context to conjugate with the universal and supreme focal point, Omega. Emersion, after emergence! (*The Human Phenomenon*, 1938–40, I, 223 E; 343–4 F)

EN-

Prefix. Expresses the sense of bringing "into a certain condition or state".

ENERGETICS

1 Praxis. Contemplation in action.
2 Science that synthesizes physico-chemical, biological and psychic energies on a human level in a generalized dynamics centered on spiritual forces.

(a) We get a hint here of a new energetics (maintenance, canalization and increase of human hopes and passions) in which physics, biology and morals would be brought together. (*The Phenomenon of Man*, 1928, IX, 96 E; 127 F)

(b) So far we have rightly been impassioned by the revelation of the mysteries hidden in the infinitely large and the infinitely small of matter. But an investigation of much greater importance for the future will be the study of psychic currents and attractions: in effect, an energetics of the spirit. Impelled by the necessity of building the unity of the world, we shall perhaps end up by seeing that the great object obscurely pursued by science is none other than the discovery of God. (*The Salvation of Mankind*, 1936, IX, 145–6 E; 187 F)

ENERGY *see also* Activance; Energy, Christic; Energy, Controlled; Energy, Human; Energy, Incorporated; Energy, Radial; Energy, Spiritualized; Energy, Tangential

Dynamism. Driving force. Basic stuff of a universe in process of personalization. Corresponds to scholastic "act". Appears in granular form, that is, develops from physico-chemical, biological and psychic centers. Energy is presence (*The Christic*, 1955, XIII, 99 E; 114 F).

(a) Energy, the third aspect of matter. Under this word, that conveys the sense of psychological effort, physics has introduced the precise formulation of a capacity for action or more exactly, for interaction. Energy is the measure of what is transferred from one atom to another in the course of their transformations . . . Never, in fact, grasped in its pure state, but always more or less granulated (even in light!), for science energy currently represents the most primitive form of the universal stuff. (*The Human Phenomenon*, 1938–40, I, 13–14 E; 36–7 F)

(b) What name should we give . . . to the physico-moral energy of personalization to which every activity revealed by the stuff of the universe is finally reduced? Only one – love: if we are to credit it with the generality and power it must assume on rising to the cosmic order. (*Sketch of a Personalistic Universe*, 1936, VI, 72 E; 90F)

ENERGY, CHRISTIC

Christ as efficient cause. Christ, through his incarnation, as mover of convergent evolution and source of supernatural or mystical dynamism. Highest form of energy.

> And so we see gradually emerging the extraordinary notion and vision of a certain universal christic energy, at once supernaturalizing and superhumanizing, in which the field of convergence necessary to explain and ensure the general and global involution of the cosmos on itself, is simultaneously materialized and personal. (*From Cosmos to Cosmogenesis*, 1951, VII, 264 E; 272 F)

ENERGY, CONTROLLED

Material energy that is increasingly controlled by human beings through science and technology with the object of organizing them for spiritual ends.

> Controlled energy is what enables us, through our own ingenuity, to dominate physically our environment by means of "artificial machines". (*Human Energy*, 1937, VI, 115 E; 146 F)

ENERGY, HUMAN

1 Human physical and psychic energy. Love is human energy par excellence.

> Since, in a universe become capable of thought, everything finally moves in and towards the personal, it is necessarily love, a sort of love, that forms and will increasingly form, in a pure state, the stuff of human energy. (*Human Energy*, 1937, VI, 145–6 E; 181 F)

2 By extension, constantly increasing element of cosmic energies that human activity employs for its own ends.

(a) By "human energy" I understand here, as a preliminary approximation, the sum of physico-chemical energies either simply incorporated or (at a higher degree of assimilation) cerebralized within the human planetary mass at a given moment – the mass in question being considered in the linked totality not only of its biological constituents but also of its artificially constructed mechanisms. (*The Activation of Human Energy*, 1953, VII, 387 E; 409 F)

(b) By human energy I understand here the constantly increasing portion of cosmic energy presently subject to the recognizable influence of centers of human activity. (*Human Energy*, 1937, VI, 115 E; 145 F)

ENERGY, INCORPORATED

Cosmic energy that is gradually concentrated and organized to form a living body.

> Incorporated energy is what the slow biological evolution of the earth has gradually accumulated and harmonized in our organism of flesh and nerves: the astonishing "natural machine" of the human body. (*Human Energy*, 1937, VI, 115 E; 145–6 F)

ENERGY, RADIAL *see also* **Energy, Tangential; Things, Inside of**
Psychic or spiritual energy. Cosmic force behind the evolutionary growth of the universe. Creative energy is interiorizing, centric and evolutionary: its progression is correlative with tangential energy. With tangential energy, constitutes an increasingly centered and increasingly complex "arrangement". Synonymous with thomist notion of the divine presence and concurrence at work in the activity of the whole universe. Equivalent to spiritual energy in Bergson and Jung (see also Émile Rideau SJ, *Teilhard de Chardin*, 79 E; 165 F).

> Briefly, the "trick" consists of distinguishing two sorts of energy: the one primary (psychic or radial energy) escaping from entropy; the other secondary (physical or tangential energy) obedient to the laws of thermodynamics – the two energies not being directly transformable into each other but mutually interdependent on each other in function and evolution (the radial increasing with the arrangement of the tangential and the tangential only becoming arranged when activated by the radial). (*The Singularities of the Human Species*, 1954, II, 265 E; 263 F)

ENERGY, SPIRITUAL *see also* **Energy, Radial; Energy, Spiritualized; Evolution, Creative**

(a) No concept is more familiar to us than spiritual energy . . . the objectivity of the effort and work of the psyche is so certain that the whole of ethics resides in it . . . There is no doubt that material energy and spiritual energy hold together and are prolonged by something. Ultimately, somehow or other there must be only a single energy at play in the world. (*The Human Phenomenon*, 1938–40, I, 28–9 E; 59–60 F)

(b) The possible increases of total spiritual energy derive, strictly speaking, from what Bergson called "creative evolution". (*Human Energy*, 1937, VI, 136 E; 170 F)

ENERGY, SPIRITUALIZED *see also* **Energy Radial; Energy, Spiritual**
Energy that, through complexification, emerges across the human threshold in the specific form of freedom, reflection and love. Electrifies the world with human intentions.

> Spiritualized energy . . . is what, localized in the immanent zones of our free activity, forms the stuff of our intellections, affections and volitions: probably an imponderable energy but an energy wholly real since it brings about a reflective and passionate control of things and their relationships. (*Human Energy*, 1937, VI, 115 E; 146 F)

ENERGY, TANGENTIAL *see also* **Energy, Radial; Entropy; Neguentropy; Things, Outside of; Transience**
Physical energy. Thermodynamic energy (subject to the second law of

thermodynamics) that determines purely external relationships between material elements, relationships that are correlative with radial energy that ultimately conditions them. Second law of thermodynamics = law of dissipation and dissociation (see also Émile Rideau SJ, *Teilhard de Chardin*, 79 E; 165 F).

> We shall assume that all energy is essentially psychic. But we shall add that in each individual element this fundamental energy is divided into two distinct components: a tangential energy making the element interdependent with all elements of the same order in the universe as itself (that is, of the same complexity and same "centricity"); and a radial energy attracting the element in the direction of an ever more complex and centered state, toward what is ahead. (*The Human Phenomenon*, 1938–40, I, 30 E; 62 F)

ENFOLDING *see also* **Arrangement; Counter-evolution; Involution; Spiral**

Teilhard uses "enfolding" or coiling to encapsulate what he sees as an essential feature of evolution. Complements traditional idea of unfolding. Process of centration or turning inwards towards a center. Contains strong echoes of Jung's individuation process. Teilhard uses the image of the enfolding spiral to illustrate the idea of a cone of convergence towards a summit. Teilhardian "keyword".

(a) One does not need to be a great prophet to affirm that within two or three generations the notion of a psychic enfolding of the earth upon itself within some "space of complexity" will be as widely used and accepted by our successors as the idea of its mechanical movement "around the sun" was accepted by us. (*Does Mankind Move Biologically Upon Itself?*, 1949, V, 259 E; 335–6 F)

(b) In the perspectives of cosmic enfolding, consciousness not only becomes co-extensive with the universe, but the universe, in the form of thought, falls into equilibrium and consistency on a supreme pole of interiorization. (*The Human Phenomenon*, 1938–40, I, 223 E; 344 F)

ENTITIES, PAIRED

Irreducible terms subject to correlative variations (being-union, spirit-matter, universal-personal, etc.).

> Rather than looking as I have done previously . . . at esse as further definable by unire (or by uniri), it might be better to consider the two notions of being and union (or, if we like, of moving body and movement) as forming a natural pair whose two terms (equally primitive and basically irreducible) are ontologically inseparable (like the two surfaces of one and the same plane) and obliged to vary simultaneously in the same direction. (*My Fundamental Vision*, 1948, XI, 207–8 E; 222 F)

ENTROPY *see also* **Neguentropy**

Involution. In evolutionary terms, disorder, disorganization and probability as opposed to order, organization and improbability. Measurement of gradual dissipation of physical energy, inevitable over time in any closed thermodynamic system. Ultimately reducing the universe to a mean state of diffuse agitation in which all exchange of useful energy ceases.

> By virtue of the statistical laws of probability . . . everything around us appears to be descending towards the death of matter; everything except life. (*The Movements of Life*, 1928, III, 149 E; 209 F)

ENTROPY, NEGATIVE *see* **Neguentropy**

EPIPHANY

> In the final analysis the immense enchantment of the divine milieu owes its true value to the divine–human contact revealed in the epiphany of Jesus. If we suppress the historical reality of Christ, the divine omnipresence that intoxicates us becomes like all the other dreams of metaphysics: uncertain, vague, conventional . . . (*The Divine Milieu*, 1926–7, 1932, IV, 104–5 E; 140 F)

EQUILIBRIUM, MORALITY OF *see also* **Ethics; Movement, Morality of**

Closed morality. Static morality expressed in legal and social relationships. Limits the scope of christian virtues.

(a) Morality arose largely as an empirical defense of the individual and society . . . Morality has been principally understood up to now as a fixed system of rights and duties intended to establish a static equilibrium between individuals . . . This concept . . . has been radically transformed if one recognizes that the human on earth is no more than an element intended to complete itself cosmically in a higher consciousness in process of formation . . . Hitherto the moralist was a jurist – or an equilibrist . . . Henceforth the highest morality is what will best develop the natural phenomenon to its upper limits. (*The Phenomenon of Spirituality*, 1937, VI, 106 E; 131–2 F)

(b) A morality of equilibrium can logically be agnostic and absorbed in the possession of the present moment. A morality of movement is necessarily focussed on the future. (*The Phenomenon of Spirituality*, 1937, VI, 109 E; 135 F)

ETHICS, TEILHARDIAN *see also* **Anthropogenesis; Equilibrium, Morality of; Movement, Morality of; Obligation; Responsibility; Spirit, Atomism of**

Open system of morality. Energetic morality. Morality integrated in a dynamic system providing virtues and values with evolutionary and energetic meaning.

(a) To be more is first of all to know more. (*Note on Progress*, 1920, V, 19 E; 31 F)

(b) We cannot fail to recognize . . . the basis for a new ethics of the earth . . . A sort of generalized ultra-responsibility . . . would appear to be the most pronounced moral characteristic of the ultra-human towards which, by cosmic necessity, we are drifting whether we like it or not. (*The Evolution of Responsibility*, 1950, VII, 212–13 E; 218–19 F)

(c) No concept is more familiar to us than spiritual energy . . . the objectivity of the effort and work of the psyche is so certain that the whole of ethics resides in it. (*The Human Phenomenon*, 1938–40, I, 28 E; 59–60 F)

(d) To admit, in fact, that a combination of races and peoples is the event biologically awaited in order for a new and higher development of consciousness to take place on earth, is at the same time to define . . . an international ethics. (*Natural Human Units*, 1939, III, 211 E; 296 F)

(e) Charity reappears, at the head of the most modern, the most scientifically satisfying of moral systems, once, having been transposed into a universe in process of being spiritually drawn closer together, it automatically becomes dynamic . . . The measure of an ethics is its ability to flower in mysticism. (*The Atomism of Spirit*, 1941, VII, 51, 53–4 E; 58, 60 F)

(f) The most traditional human morality takes on new form, new coherence and new urgency . . . The main lines of a whole new philosophy of life, a whole new ethics and a whole new mysticism are correlatively picked out by the same shaft of light. (*The Atomism of Spirit*, 1941, VII, 49 E; 56 F)

(g) We now see the rules of ethics, hitherto considered as more or less freely superposed on the laws of biology, showing themselves, not metaphorically but literally, to be the very condition of survival for the human race. (*The Human Rebound of Evolution*, 1947, V, 203–4 E; 261–2 F)

(h) The moralization of invention . . . The planetary pressures that are compelling us to unite can only operate effectively under certain psychic conditions . . . that merely recall and re-express the broad lines of a traditional and empirical ethics. (*The Human Rebound of Evolution*, 1947, V, 202 E; 260 F)

(i) Properly speaking, there are neither sacred nor profane things, neither pure nor impure things. There is only a good direction and a bad direction: the direction of ascent, of expanding unification, of greater spiritual effort; and the direction of descent, of constricting egoism, of materializing enjoyment. (*The Evolution of Chastity*, 1934, XI, 72 E; 78 F)

EU-

Prefix. Expresses the sense of “good”, “well”, “easily”, e.g. eu-centric, eu-centrism, eucharistization, eu-collectivization, eu-complex, eu-cosmogenesis, eugenics, eu-pantheism . . .

EU-CENTRIC *see also* **Eu-centrism**

In the domain of the eu-centric, the noospheric center, omega, is not born

of the confluence of human "egos" but emerges on their organized totality like a spark leaping the gap between the transcendent face of omega . . . and the "point" of a perfectly united universe. (*Centrology*, 1944, VII, 114 E; 120 F)

EU-CENTRISM *see also* **Centricity, Fragmentary; Centricity, Phyletic; Centrogenesis; Eu-centric**

Third stage of centrogenesis. Appearance of nuclei in punctiform centers that allow the threshold of reflection to be crossed.

> Just as, at the origins of the phyletic, the closing-up on itself of a chain of segments (centration) determined the first appearance of living centers – so here, through its centric diameter (reflection) reaching zero, the living center in its turn attains the condition and dignity of a "grain of thought". (*Centrology*, 1944, VII, 109 E; 115 F)

EUCHARIST *see also* **Eucharistization; Mass, The**

Christ truly present in the bread and wine of the Eucharist. The Real Presence. The Eucharistic Christ.

> The Host, of course, is first and foremost the fragment of matter to which, through transubstantiation, the presence of the Word Incarnate "attaches itself" among us, that is, in the human zone of the universe. The center of personal energy of Christ is truly fixed in the Host. And, just as we rightly call "our body" the local center of our spiritual radiation (without necessarily meaning that our flesh is more ours than any other matter), we should say that the initial Body of Christ, his primary Body, is confined to the species of bread and wine. (*My Universe*, 1924, IX, 65 E; 93 F)

EUCHARISTIZATION *see also* **Eucharist; Mass, The**

Progressive assimilation of humanity and, through humanity, of the universe. Process indispensable to the completion of the Mystical Body of Christ.

> All the communions of our lives are no more than successive instances or episodes of one single communion, of one and the same process of christification . . . The Eucharist, considered in its totality, is none other than the expression and manifestation of the divine unifying energy applied in detail to each spiritual atom of the universe. In short, to adhere to Christ in the Eucharist is inevitably and ipso facto to be incorporated a little more each time in a christogenesis that is none other (and this, as we have seen, is the essence of the christian faith) than the soul of the universal cosmogenesis. (*Introduction to the Christian Life*, 1944, X, 166 E; 194–5 F)

EU-COMPLEX *see also* **Complexification; Complexity; Complexity-consciousness, Law of**

Biological complex at the point of evolution where it generates consciousness that is reflective and, consequently, personalized. Opposite of pseudo-complex.

The type of super-grouping, towards which the course of civilization is driving us, far from simply representing some material aggregates ("pseudo-complexes") where elementary freedoms are neutralized by the effects of large numbers or mechanized by geometric repetition, belongs on the contrary to the species of "eu-complex" or arrangement where arrangement, because it is, and to the extent that it is, a generator of consciousness, is ipso facto classified as biological in nature and value. (*Man's Place in Nature*, 1949, VIII, 101–2 E; 146–7 F)

EUGENICS

Science of conditions favorable to maintaining the quality of the human species.

(a) It is indispensable in the course of the coming centuries that a highly human form of eugenics be discovered and developed, on the personal scale. A eugenics of the individual – and it follows, a eugenics of society as well. (*The Human Phenomenon*, 1938–40, I, 202 E; 314 F)

(b) An optimum arrangement with a view to a maximum humanization of the noosphere. Hence, and above all, a fundamental preoccupation with ensuring (nutritively, educationally, selectively) a growing eugenics of the human zoological type on the surface of the earth. (*The Sense of the Species*, 1949, VII, 202 E; 209 F)

EU-PANTHEISM *see also* Pantheism; Para-pantheism; Union, Pantheism of

(a) In a slightly different form we could say that the pantheist (or cosmic) sense tends to be expressed according to one or other of the following three formulae:

(1) to become all beings (an impossible and mistaken act): para-pantheism;
(2) to become all ("eastern" monism): pseudo-pantheism;
(3) to become one with all beings ("western" monism): eu-pantheism. (*Reflections on Two Converse Forms of Spirit*, 1950, VII, 223 E; 232 F)

(b) Distinguish between the omnipresence of God and the multipresence of Christ where omni = homogeneous, common basis and multi = multiform: each contact specifically modalized = eu-pantheism . . . (Retreat, Les Moulins, 25 August 1949, in Claude Cuénot, *Nouveau Lexique*, 151)

EVIL *see also* Diminution, Passivities of; Growth, Passivities of

1 In an ontological sense, waste caused by the process of unification that falls back into the multiple.
2 In a psycho-physiological sense, passivities of growth and diminution: suffering, pain, illness.

3 In a moral sense, freedom to make mistakes.
4 In a religious sense, freedom to sin where fault is seen as a refusal of God. Dimension involving ethical, metaphysical and theological considerations. The religious sense encompasses the ontological, psycho-physiological and moral senses.

(a) True non-being, physical non-being, non-being on the threshold of being, non-being where all possible worlds converge at their base, is pure multiple, multitude . . . Non-being, pain, sin – ontological evil, perceived evil, moral evil – three aspects of the same principle of evil, reducible over an infinity of time and constantly being reborn – the multitude. (*The Struggle against the Multitude*, 1917, XII, 95, 103 E; 132, 141 F)

(b) For God, to attack the multiple is inevitably to come into conflict with evil, "the shadow of creation". (*My Fundamental Vision*, No. 29, 1948, XI, 196 E; 211 F)

EVIL, CATASTROPHIC *see also* **Evil, Evolutionary**

Non-teilhardian concept according to which fault (reflective evil) and pain (non-reflective evil) that is the consequence of fault, are caused by the sudden disarrangement, through sin, of a previously stable and harmonized world. As such, catastrophic evil does not represent the price of progress.

> In a system of cosmos . . . it was extremely difficult, if not impossible (except by intervention of an accident, itself quasi-inexplicable), to justify in our minds the existence of pain and error in the world. In a system of cosmogenesis, on the other hand, how much longer must we cry out to make ordinary "public opinion" understand that, intellectually (I do not say emotionally) speaking, not only does the problem of evil become soluble – it no longer arises? . . . Evil, a secondary effect, an inevitable by-product, of the process of a universe in evolution . . . An evil no longer catastrophic but evolutionary. (*From Cosmos to Cosmogenesis*, 1951, VII, 259–60 E; 267–8 F)

EVIL, EVOLUTIONARY *see also* **Evil, Catastrophic**

Teilhardian concept according to which fault (reflective evil) and pain (non-reflective evil) are linked to the structure of a world in becoming where being in process of unification is torn between the appeal of the one and a constantly threatening multiplicity. Evolutionary evil, through its ontological origins linked to the existence of a non-unified multitude, is a condition prior to the appearance of the human.

> In a system of cosmogenesis . . . how much longer must we cry out to make ordinary "public opinion" understand that, intellectually (I do not say emotionally) speaking, not only does the problem of evil become soluble – it no longer arises? . . . Evil, a secondary effect, an inevitable by-product, of the process of a universe in evolution . . . An evil no longer catastrophic but evolutionary. (*From Cosmos to Cosmogenesis*, 1951, VII, 259–60 E; 267–8 F)

EVOLUTION *see also* **Curve; Entropy; Involution; Spiral**

1 In a scientific and phenomenal sense, law of the succession and transformation of phenomena in time (general evolution at every level of being). Life.

At present, human knowledge is developing entirely under the aegis of an evolution seen as the prime property of an experiential real; so true is this that we are unable to conceive of anything that does not satisfy from the outset the conditions of a universe in course of transformation. (*Christology and Evolution*, 1933, X, 78 E; 97 F)

2 In a cognoscitive sense, dimension of all thought. The only way of knowing the real.

(a) At this level of generality (where all experiential reality is part of a process of birth in the universe), evolution has long ceased to be a "hypothesis" but has become a general condition of knowledge (a further dimension) to which henceforth all hypotheses must conform. (*The Singularities of the Human Species*, 1954, II, 211 n.1 E; 298 n.2 F)

(b) No longer a hypothesis but a condition to which henceforth all hypotheses must conform. The expression in our minds of the shift from a state of "cosmos" to a state of "cosmogenesis". (*Evolution of the Idea of Evolution*, 1950, III, 246 E; 348 F)

(c) Evolution is a light illuminating all facts, a curve that every line must follow. (*The Human Phenomenon*, 1938–40, I, 152 E; 242 F)

3 In a total and ontological sense, phenomenal and temporal dimension of a transphenomenal action, namely, divine creation. Continuous creation (*creatio continua*).

From this point of view . . . evolution takes on its true form in our minds and hearts. Evolution . . . is not "creative" as science once believed but is the expression, in time and space, of our experience of creation. (*Man's Place in the Universe*, 1942, III, 231 E; 323–4 F)

Teilhard does not claim to have put forward a new evolutionary theory. His basic postulate is that the creation of the world was not an event that happened in some dim past but is an ongoing process. The cosmos is a product of cosmogenesis; cosmogenesis has transcended itself to produce life and initiate biogenesis; biogenesis has transcended itself to produce man and initiate noogenesis. These are scientifically demonstrable facts. From here on Teilhard proceeds by inference on the basis of faith. Noogenesis will be fulfilled in convergence in omega. (Theodosius Dobzhansky, *Teilhard Review*, The British Teilhard Association, V.2, 69)

EVOLUTION, CREATIVE *see also* **Creation, Evolutionary**

Expression used in the title of the book (1907) by Henri Bergson (1859–1941). Teilhard speaks of evolutionary creation rather than creative evolution.

EVOLUTION, HUMAN REBOUND OF

Threshold where a renewal of the evolutionary movement, that previously appeared to have been halted, and a sort of electrification resulting from consciousness "born" of complexity are now able to react on this same complexity in order to bring about a new breakthrough ahead of the human and humanity.

> What is now being shaped within planetized humanity is essentially, unless I am mistaken, a rebound of evolution on itself. We have all heard about these projectiles, real or imaginary, whose motion is periodically renewed thanks to the firing of one stage after another. Is not life at this moment preparing a similar process in order to make the final leap? . . . Really this is life setting course again on a second adventure from the springboard it established when forming humanity. (*The Formation of the Noosphere*, 1947, V, 176–7 E; 223–4 F)

EVOLUTION, PAIN OF

Positive effort and suffering at the basis of all progress.

> A dangerous double moral crisis is being initiated by the mental phenomenon of reflection. A crisis of emancipation, first and foremost, derived from the birth of freedom. But also a crisis of panic, linked to the psychological shock of being brutally awoken in the night. And it is precisely and only this form (the apprehensive) assumed in human consciousness by what we might call the "evil" or, at least, the "pain of evolution" to which I shall address myself here. (*A Phenomenon of Counter-Evolution in Human Biology*, 1949, VII, 183 E; 189 F)

EVOLUTIONARY *see also* **Evil, Evolutionary; Evolution**

Characteristic of evolution.

> One can imagine certain abnormal individuals exercising their freedom of choice of refusal of denying (to their loss) the hope of the "evolutionary vortex". But such denials can only be regarded as a waste. (*The Singularities of the Human Species*, 1954, II, 245 E; 339 F)

EVOLUTIONISM *see also* **Evolution**

Generalized concept of evolution.

> Scientific evolutionism is not only a hypothesis for the benefit of zoologists but also a key anyone can use to penetrate any area of the past – the key to the universal real. (*Basis and Foundations of the Idea of Evolution*, 1926, III, 127 E; 178 F)

EVOLVER *see also* **Christ-the-Evolver**

Agent of evolution.

> To say that Christ is the term and mover of evolution – to say that he reveals

himself as "evolver" – is to recognize implicitly that he becomes attainable in and through the whole process of evolution. (*Super-Humanity, Super-Christ, Super-Charity*, 1943, IX, 167 E; 212 F)

EXCENTRATION *see also* Decentration

Second stage of the existential dialectic of happiness. Concept that adds a nuance of drama and sorrow to the notion of extraction or separation.

> Ultimately, the world can only be joined to you, Lord, by a sort of inversion, turning inside out or excentration where not only individual success but even the appearance of all human progress must be temporarily overshadowed. (*The Mass on the World*, 1923, XIII, 129 E; 151 F)

EXERCISES, SPIRITUAL

Ignatian Exercises. Focus of Teilhard's spiritual life. Like every member of the Society of Jesus, Teilhard took eight to ten days each year to make *The Spiritual Exercises* (approved by Pope Paul III in 1548). He constantly reflects and asks (with St. Ignatius): "What have I done for Christ? What am I doing for Christ? What should I do for Christ?" (Exercise 53). Over the years Teilhard became increasingly concerned with bringing together the mystical vision of St. Ignatius and the new world view of "a universe both cosmically and christically convergent" (letter to Auguste Valensin SJ, 8 August 1950, *Lettres intimes*, 392; see also the dynamic Omega of Teilhard and the static Omega of the Spiritual Exercises, Jacques Laberge SJ, *Pierre Teilhard de Chardin et Ignace de Loyola*, 193).

(a) More and more, "making a retreat" is becoming simply a return to the presence of God. I just cannot understand how people can still be satisfied with (or even carried away by) the Exercises. Their scheme is splendid but the "cosmology" (and hence the christology) is so childish that it literally stifles me from beginning to end. What we need is a complete transposition of the Ignatian scheme into an organic type of universe (a universe in genesis) as we now understand it. I think such a transposition is perfectly possible (and I shall perhaps try to sketch it out one day) – but who would want to consider it? This reflects in miniature the whole drama of the Church today. (Letter to Pierre Leroy SJ, 29 August 1950, *Lettres familières*, 72; *Letters from My Friend*, 63)

(b) No longer "The Exercises" but (very humbly and because I am compelled from within) "My Exercises". The result of others, naturally. But with "The" in addition. The neo-foundation, I repeat, Neo-Christ, Neo-Cross – the new Heart. And this, I repeat, in complete submission to the Church; since the Neo-Christ can only be a Trans-Christ. (Retreat, Purchase, 23 June 1952, in Claude Cuénot, *Nouveau Lexique*, 134)

(c) I cannot avoid noting once again (and more profoundly than ever) the abyss that little by little has opened up between my religious vision of the world and that of the Exercises (in the mould into which those in high places still think we can fit!). "Abyss" not of contradiction but of expansion. The circle that has become a sphere . . . (Letter to Pierre Leroy SJ, 27 June 1952, *Lettres familières*, 145; *Letters from My Friend*, 133)

(d) Have you ever realized the extent to which the Exercises (Foundation, Sin, Reign, Standards – everything except the "*ad amorem*") must absolutely and can magnificently be transposed into a universe in genesis where, besides (at the base of) the "supernatural", we must make a place for the "ultra-human"? (Letter to Pierre Leroy SJ, 4 April 1955, *Lettres familières*, 247; *Letters from My Friend*, 215)

(e) Ignatius did not see what could (and must, in fact,) "equip" his Society was not static unity (conformism) in doctrine but doctrinal "pioneerism" following "the christic axis" and "the pleromic axis" (static orthodoxy versus dynamic orthodoxy . . .). (Journal, 13 March 1955, in Jacques Laberge SJ, *Pierre Teilhard de Chardin et Ignace de Loyola*, 200–1)

EXPANSION, SOCIALIZATION OF *see also* **Compression, Socialization of**

First phase of the process of socialization that consists of a geographical expansion through dispersion from a human upsurge that coincides with the phenomenal origins of the human.

From its origins to thc present day, humanity, while gathering and organizing inchoatively on itself, certainly passed through a period of geographical expansion, in the course of which its first concern was to multiply and occupy the earth. And it is only recently, "once the line was crossed", that the first symptoms appeared in the world of a definitive and global withdrawal of the thinking mass within a higher hemisphere where it can only advance by contracting and concentrating under the influence of time. Socialization of expansion, reversing itself, to culminate in socialization of compression. (*Man's Place in Nature*, 1949, VIII, 81–2 E; 118–19 F)

EXTENSION, PANTHEISM OF *see also* **Détente, Pantheism of; Para-pantheism**

Aesthetic attitude that seeks God in the present and in an indefinable multiplicity of experiences.

EXTRAPOLATION

Means whereby a curve is constructed, with the aid of positive data, beyond what can be observed. Teilhard uses the data of evolution to predict the future of the evolutionary process. His extrapolation is grounded in the unity and coherence of the universe.

The fact that love is growing instead of diminishing quite naturally finds its explanation – and its extrapolation – in the hypothesis of a universe in process of personalization. Evolution does not cease in the human individual but continues towards a more perfect concentration linked to further differentiation through union . . . (*Sketch of a Personalistic Universe*, 1936, VI, 74 E; 92–3 F)

EXTRINSICAL *see also* **Extrinsicism; Marxism; Opium; Terrenism**

In fact, every conversation I have ever had with communist intellectuals has left me with the definite impression that marxist atheism, far from being absolute, merely rejects the "extrinsical" sort of God, a "Deus ex machina", whose existence can only harm the dignity of the universe and weaken the source of human effort: a "pseudo God", in short, no one today (and not least the christians) wants any longer. (*The Heart of the Problem*, 1949, V, 266–7 E; 346 F)

EXTRINSICISM *see also* **Juridicism**

Exteriority between every manifestation of the real that excludes all organicity. Hence, the refusal to consider any link of immanence between God and his creation. Mode of thought condemned by Teilhard because it introduces a radical discontinuity between levels of being and even between individual beings.

General subject: extrinsicism – separatist tendency of beings (excess contrary to pantheism). Visible expression everywhere; *verbi gratia*: "arbitrary" creation; creation = efficiency (God and kosmos linked *ex voluntate*); body and soul = considered as strangers (a dead-weight and a captive); nature and supernature = simply occasional links. (Journal, 16 December 1918, *Journal*, 379)
Verbi gratia = for example
Ex voluntate = by an act of pure will

F1 *see also* **Emersion**

Geometric symbol. Complexity. Focus of arrangement. Material focus of being. Technological infrastructure of human collectivity.

> Earlier . . . I noted that every natural unit produced by cosmic involution can be represented symbolically by an ellipse constructed around two conjugated foci: one, F1, of material or technological arrangement; the other, F2, of consciousness (that can only appear at the level of life). (*My Fundamental Vision*, No. 18, 1948, XI, 183 E; 199 F; see also II, 241 fig. 20 E; 334 fig. 5 F)

F2 *see also* **Emersion; Energy, Radial**

Geometric symbol. Consciousness. Focus of consciousness. Spiritual focus of being. Technological interiorization within the human collectivity under the influence of radial energy.

> Every individual cosmic particle can be represented symbolically in our experience as an ellipse constructed around two foci of unequal and variable intensity: one, F1, of material arrangement, the other, F2, of psychism; F2 (consciousness) appearing and increasing initially as a function of F1 (complexity) but soon revealing a persistent tendency to react constructively on F1, in order to super-complexify it and become increasingly individualized itself. (*My Fundamental Vision*, No. 2, 1948, XI, 167 n.3 E; 184 n.2 F; *see also The Singularities of the Human Species*, II, 241 fig. 20 E; 334 fig. 5 F)

FAITH

1 Act of spirit. Anticipates the realization of increasingly newer and higher syntheses based on earlier syntheses.

> Christian philosophy does not exist in the sense of two opposed entities. Faith is only born of faith. Reason is not on one side and faith on the other. More and more elevated acts of faith. The world has a sense, a direction – a sense of spirit – a sense achieved through unification – adherence to christianity. Philosophy is pre-christian but adopts the attitude of a faith. (Lecture, Catholic Union of the Sick, 1933, notes by Claude Cuénot, in idem, *Teilhard de Chardin*, 141 E; 175 F)

2 Defined by Teilhard in a traditional christian sense in terms of its energetic, unitive and creative dimension.

> Faith, as we understand it here, is certainly not only intellectual adherence

to christian dogma. In a much deeper sense, it is belief in God filled with confidence in his beneficial strength that knowledge of this divine being can arouse in us. It is the practical conviction that the universe, in the hands of the Creator, continues to be the clay in which he shapes numerous possibilities according to his will. In a word, it is the evangelical faith of which we can say that no virtue, not even charity, has been more strongly recommended by the Savior. (*The Divine Milieu*, 1926–7, 1932, IV, 126 E; 168–9 F)

FALL *see* **Creation**

FATHERS, GREEK

Major influence on Teilhard, e.g. St. Gregory of Nyssa (331–95), St. Irenæus of Lyon (*c.*130–*c.*200). The "Age of the Fathers" did not come to an end in the fifth century: many later writers are also "Fathers". They remain living witnesses. Our own age might well produce a new St. Basil the Great (*c.*329–79) or a new St. Athanasius (*c.*296–373).

(a) What would happen if one were to try, following a line already suggested by the Greek Fathers, to transpose the data of revelation into a universe in movement? Such an idea is engaging the attention of a number (an ever increasing number) of catholic thinkers today. (*Catholicism and Science*, 1946, IX, 189 E; 239 F)

(b) Was it not St. Thomas who, comparing (what today we should call the fixist) perspective of Latin Fathers like St. Gregory to the "evolutionary" perspective of the Greek Fathers and St. Augustine, said of the latter, "Magis placet" (II Sent, d.12; g.I.a.I)? Let us be glad to strengthen our minds by contact with this great thought! (*What Should We Think of Transformism?*, 1930, III, 154 E; 217 F)
Magis placet = it is more pleasing

(c) The full mystery of baptism is no longer to cleanse but (as the Greek Fathers understood so well) to plunge into the fire of the purifying struggle "for being". No longer the shadow, but the passion of the Cross. (*Christology and Evolution*, 1933, X, 85 E; 103–4 F)

FATHERS, LATIN *see also* **Fathers, Greek**

FEAR, EXISTENTIAL *see also* **Anguish**

Fear linked to the human condition and caused by whatever transcends quantitatively the human being both inside and outside (both the psychic and cosmic abysses).

FEMININE, THE *see also* **Dyad**

Universalization of the singular being who is woman without losing anything of being personal. As intermediary between the human and the cosmos, the feminine stimulates the cosmic energy in the masculine with whom she forms the dyad – the unity of two persons in a single whole that must be super-centered on God. The dialectics of the feminine and the masculine is an essential agent of the unitive energy of love.

(a) Woman is, for man, the symbol and the personification of all the complementarities expected of the universe. The theoretical and practical problem of the fulfillment of knowledge has found its natural "climate" in the sublimation of love. At the term of the spiritual power of matter is found the spiritual power of the flesh and the feminine. (*The Evolution of Chastity*, 1934, XI, 70 E; 76 F)

(b) None of us can dispense with the feminine any more than we can dispense with light, oxygen or vitamins. (*The Feminine or the Unitive*, 1950, XIII, 59 E; 72 F)

(c) Let us not repeat, through ideology or sentimentality . . . the error of feminism or democracy at the beginning. Woman is not man: and it is precisely for this reason that man cannot do without woman. A mechanic is not an athlete, or an artist, or a financier . . . These inequalities, that against all the evidence we sometimes try to deny, may appear damaging as long as the elements are regarded statistically in isolation. Observed, however, from the point of view of their essential complementarity, they become acceptable, honorable and even desirable. (*Natural Human Units*, 1939, III, 212 E; 297–8 F)

FINALITY *see also* **Movement, Neo-anthropocentrism of; Orthogenesis; Universe, Personal**

1 In the narrowest sense, causality of the reflective psyche by which the human acts inventively in striving towards a given end by the intervention of given means – ends and means always being open to revision for internal reasons. Reflection combines the forward projection of a plan with the unforeseeability of invention.

Born under the appearance and sign of chance it is through the low subjection of reflective finality that life can hope to lift itself, through auto-evolution, henceforth in the dual direction of the highest complexity and the greatest complexity. (*The Human Rebound of Evolution*, 1947, V, 201 E; 258 F)

2 In a wider sense, characteristic of the human plan as opposed to the biological plan governed by orthogenesis. Hence the use of the term "finality" rather than "finalism" (where the latter encapsulates the biological and physico-chemical levels and is derived, in the teilhardian view, from a static anthropomorphism).

If the neo-darwinians are right (as is possible and even probable) in claiming that in the pre-human zones of life they see nothing but the play of chance in the progress of the organized world – on the other hand, beginning with the human, it is the neo-lamarckians who now have the advantage since, starting from this level, the forces of internal arrangement begin to show themselves distinctly in the process of evolution. That amounts to saying that biological finality . . . is not perceptible everywhere but only makes itself felt in the world above certain levels. (*The Human Rebound of Evolution*, 1947, V, 200 E; 257–8 F)

3 In the most general sense, finality encompasses the entire universe within the perspectives of an evolutionary neo-anthropocentrism or neo-anthropocentrism of movement where the finalized action of human beings takes hold of a universe in process of personalization.

Undoubtedly, if we look at the universe as it is now we have to admit that it is oriented towards a determined final end: without such an end it would collapse in dust. But we must understand that "*secundum signa rationis (ratiocinatæ)*", this finality that covers, coordinates and animates the general movement of things, is a descending creative influence, a gradually defined synthesizing process. (*Forma Christi*, 1918, XII, 251–2 E; 367 F)

FIRE *see also* **Heart, Sacred**

Symbol of the divine. Metaphor frequently used by Teilhard. Fire occupies a central place in his mysticism. He speaks of himself and the universe as "incandescent". The Teilhard family motto is taken from Virgil, *Æneid*, VI. 730: "*Igneus est illis vigor et celestis origo*" ("Their strength is of fire and their origin of heaven") (see also Henri de Lubac SJ, *Lettres intimes*, 86–7 n.6).

(a) Fire, source of being: we are held fast by the tenacious illusion that it comes from the depths of the earth and that its flames burn progressively ever brighter along the trail of life . . . In the beginning was power, intelligent, loving, energizing. In the beginning was the word, supremely capable of mastering and molding whatever matter might come into being. In the beginning was not coldness and darkness but fire . . . Blazing spirit, fire, fundamental and personal, real term of a union a thousand times more beautiful and desirable than the destructive fusion imagined by no matter what pantheism . . . Fire once more, has penetrated the earth.. without tremor, without thunderbolt, the flame has illuminated all from within. (*The Mass on the World*, 1923, XIII, 121–3 E; 143–5 F)

(b) The full mystery of baptism is no longer to cleanse but (as the Greek Fathers understood so well) to plunge into the fire of the purifying struggle "for being". No longer the shadow, but the passion of the Cross. (*Christology and Evolution*, 1933, X, 85 E; 103–4 F)

(c) I was still not yet "in theology" when, through and in the symbol of the "sacred heart", the divine had already taken on for me the form, consistence and properties of an energy, of a fire, in other words, having become capable of spreading itself everywhere and of metamorphosing itself into no matter what, it henceforth found itself, by virtue of being universalizable, capable of forcing itself into and amorizing the cosmic milieu . . . (*The Heart of Matter*, XIII, 1950, 44 E; 55 F)

(d) The day will come when, after harnessing space, the winds, the tides and gravitation, we shall harness for God the energies of love. And on that day, for the second time in the history of the world, we shall have discovered fire. (*The Evolution of Chastity*, 1934, XI, 86–7 E; 92 F)

FIXISM *see also* **Logicalism; Systematics**

Hypothesis, rejected by Teilhard, according to which there has been no evolution of living species since creation (that was an act completed in time). Requires the immediate creation by God of independent species. Solidarity between species is based on a logical and divine plan rather than a chronological and phenomenal law of birth.

Faced with the important fact of "natural" distribution (geographical, morphological, temporal) of living species, the three major objections of the fixists to transformism are:

(a) impossibility of varying artificially even the smallest species identified by systematics;
(b) impossibility of paleontology tracing the precise origins of many evolutionary branches;
(c) unchanging existence of certain living species throughout geological time . . .

These objections, in our view, simply disappear, no longer exist . . . The "major objections" of fixism express no more than the characteristics or weaknesses that are to be found throughout the history of science. (*What Should We Think of Transformism?*, 1930, III, 153 E; 215–16 F)

FORMS, ORTHOGENESIS OF *see also* **Canalization, Mutation by; Instrumental differentiation, Evolution by; Phyletization**

Phenomenon of morphological accentuation capable of producing a reshaping of the entire organism.

Orthogenesis of forms. Or morphological accentuation of animal species . . . it would seem, at first sight, that in the proliferation of neighboring forms, each independent of the other, orthogenesis vanishes, like an illusion, from the moment we try to look at it closely. (*A Defense of Orthogenesis*, 1955, III, 271 E; 387 F)

FORWARD *see also* **Ahead**

Retreat seen as a deepening of contact with God-Christ Omega . . .

(1) Contact of existence,
(2) Contact of dependence (consistence),
(3) Contact of human involution . . . forward,
(4) Contact of christian involution . . . upward,
(5) Contact of interiorization . . . inward,

(6) Contact of excentration . . . pan-communion.
(Retreat, Les Moulins, 30 August–7 September 1948, 4th day, in Claude Cuénot, *Nouveau Lexique*, 95)

FRANCIS, ST. *see also* **Ignatius, St.**

Francesco Bernardone, Francis of Assisi (1181–1226), patron saint of ecologists (1980). Founder of the Friars Minor (1210).

(a) St. Bruno, says Fr. [Léonce] de Grandmaison, wanted to imitate Christ the Solitary; St. Francis saw and wanted to bring about the reign of Christ the Poor; St. Dominic, the True Christ; St. Ignatius, Christ the Chief. Who then will see and find the means of bringing about the reign of Christ Alpha and Omega, the Christ of St. Paul, the Universal Christ? May I be, in my or through my death, the humble precursor of such a man! (Journal(?), 19 August 1920, *Lettres intimes*, 50)

(b) I dream of a new St. Francis or a new St. Ignatius who would show us the new sort of christian life (both more involved in and more detached from the world) of which we have need. (Letter to Auguste Valensin SJ, 21 June 1920, *Lettres intimes*, 77)

FRONT, HUMAN *see also* **Futurism; Personalism; Universalism**

Teilhardian notion inspired by the popular fronts of the nineteen-thirties.

(a) Why do not we in our turn who now have all the space and time to develop the only possible liberty, the only possible equality and the only possible fraternity (namely, those born of collaboration in a common task), rise together for the rights of the world in the name (not so abstract as one might believe) of the future, the universal and the person? (*The Salvation of Mankind*, 1936, IX, 143 E; 185 F)

(b) This immanent elaboration of total human energy is mechanically prepared by the play of the most immediate necessities of life. Historical materialism, Marx would say. In order to obtain the collective results of organization and discovery necessary for their subsistence, the elementary thinking units are automatically led to form a linked operational group: a "human front". (*Human Energy*, 1937, VI, 136 E; 170–1 F)

FUTURISM *see also* **Personalism; Universalism**

One of the three teilhardian pillars of the future. Three elements of faith in the teilhardian human front of liberation.

(a) Futurism (by which we understand the existence of an unlimited sphere of perfection and discovery), universalism and personalism . . . the three unshakable axes on which our faith in human effort can and must depend with complete assurance. Futurism, universalism and personalism: the three pillars of the future. (*The Salvation of Mankind*, 1936, IX, 137 E; 178 F)

(b) Christianity is universalist . . . Christianity is supremely futurist . . . Christianity is specifically personalist . . . (*The Salvation of Mankind*, 1936, IX, 149–50 E; 191 F)

G

GENESIS

Orientated, generally convergent process.

GEOBIOLOGY

Science that brings together the geosphere and the biosphere to link the living layer of the earth with its physico-chemical evolution.

> Defined as the "science of the biosphere", geobiology immediately stands on its own . . .
> (1) Study, first, of organic links of every description recognizable between living creatures, considered as forming in their totality a single closed system.
> (2) Study, next, of physico-chemical links binding the birth and development of this "closed living layer" to planetary history. ("Géobiologie et Geobiologia, *Geobiologia*, 1943, in Claude Cuénot, *Teilhard de Chardin*, 229 F, 283 F)

GLOW, GOLDEN *see also* **Christ, Heart of; Heart, Sacred**

Luminous fringe revealing the divinity of Christ (epiphany and transfiguration) and the active presence of the divine milieu.

(a) The Golden Glow: the glimmer of fire – love (attraction) and evolution of love – the all, centered and non-diffused (expansion). (Journal, 24 September 1948, in Claude Cuénot, *Nouveau Lexique*, 96)

(b) The Heart of Christ at the heart of matter. The "Golden Glow" as I like to say in English. (Letter to Jeanne-Marie Mortier, 30 October 1948, *Lettres à Jeanne Mortier*, 48)

GOD

Aquinas postulates the existence of God from order in the cosmos. Teilhard postulates his existence from order in cosmogenesis (see also Bruno de Solages OCarm, *Teilhard de Chardin*, 240–54).

(a) I believe we must lay down . . . two conditions that the God we are seeking must structurally satisfy if he is to be capable of supporting and directing the spiritual phenomenon . . .

- a God of cosmic synthesis in whom we may be aware of advancing and uniting ourselves by the spiritual transformation of the powers of matter;
- a supremely personal God in whom the more we are absorbed the more we can distinguish ourselves. (*The Phenomenon of Spirituality*, 1937, VI, 109 E; 135–6 F)

(b) God is a person . . . We must think of him as we think of a person. A God who is not a person would not be God. (Conversation with Pierre Leroy SJ, in Claude Cuénot, *Teilhard de Chardin*, 244 E; 299 F)

(c) The unification of beings in God cannot be conceived as operating by fusion (God being born by the welding together of the elements of the world or on the contrary by absorbing them in himself) but by differentiating synthesis (the elements of the world becoming more themselves, the more they converge on God). (*Christian Life*, 1944, X, 171 E; 199 F)

GOD, NEW

Expression representing, not the renewal of God himself, but the new dimensions that are perceived by the human spirit in line with its developing world-view.

> For millions and millions of believers (among the most aware human beings) Christ, since he appeared, has never ceased re-emerging after each crisis of history with greater presence, greater urgency and greater penetration than ever before. What does he still lack if he is to show himself once again to our modern world as the "new God" we expect? Two things in my view; and two things only. First, in a universe where we can no longer seriously consider thought an exclusively terrestrial phenomenon, he can no longer be restricted *constitutionally* in his operation to a simple "redemption" of our planet. And second, in a universe where we can now see everything co-reflected along a single axis, he can no longer be offered for our worship (as the result of a subtle and pernicious confusion between "super-natural" and "extra-natural") as a distinct and rival summit to the summit to which the biologically extended path of anthropogenesis is leading us. (*The God of Evolution*, 1953, X, 241–2 E; 289–90 F)

GOD-ABOVE *see also* Above

God above the world. The essentially transcendent God. The traditional face of God (the God of the Genesis narratives, the God of judaism, christianity and islam . . .) that is by no means incompatible with the newly-discovered face of God in a period of maturation and convergence.

> A God-ahead suddenly appearing transversally to the traditional God-above in such a way that henceforth, unless we merge the two images in one, we shall no longer be able to worship fully. A new faith comprising an ascensional faith towards the transcendent and a propulsive faith towards the immanent . . . this is what, under threat of failure, the world now desperately awaits. (*The Heart of Matter*, 1950, XIII, 53 E; 65 F)

GOD-AHEAD *see also* Ahead

God ahead of the world. God the Prime Mover Ahead, the Mover, the Collector, the Consolidator. The immanent God. The new face of God as the term of evolution, the world and human effort. God who has acted from the very beginning without his presence being capable of being wholly revealed before the end of time.

> The synthesis of the (christian) "God-above" and the (marxist) "God-ahead" is the only God we can worship from now on "in spirit and in truth". (Letter, May–June 1952, in Claude Cuénot, *Teilhard de Chardin*, 369 E; 449 F)

GOD-THE-EVOLVER *see also* Cosmogenesis, God of

> In this new God-the-Evolver, surging up at the very heart of the old God-the-Worker, primordial transcendence must, of course, and in the first place, be maintained at all costs (and of cosmic necessity). (*From Cosmos to Cosmogenesis*, 1951, VII, 262 E; 271 F)

GOD-THE-WORKER *see* God-the-Evolver

GRACE

Free and unmerited favor of God.

> Organic nature of grace. Under the unifying effect of divine love, the spiritual elements of the world (the "souls") are raised to a higher state of life. They are "super-humanized". Consequently, the state of union with God is much more than some legalistic justification linked to an extrinsic increase in divine benevolence. From the christian, catholic and realist point of view, grace represents a physical super-creation. It causes us to climb another step on the ladder of cosmic evolution. In other words, it is a stuff that is strictly biological. (*Introduction to the Christian Life*, 1944, X, 152–3 E; 180–1 F)

GRADIENT

Axis on which a given property varies continuously (where x'Ox = axis of morphological differentiation).

> Let us try . . . to arrange diagrammatically the various cerebralized phyla we know along the radii of a semi-circle: each phyletic radius making with the diameter x'Ox an angle proportionate to its spread (its gradient) of cerebration. (*The Singularities of the Human Species*, 1954, II, 220 E; 310 F)

GRANULATION *see also* Corpusculization

Process of forming granules or grains.

> As the elements of the phyletic chain, as a direct result of corpusculization, increase in interiority and liberty, so the "temptation" inevitably grows stronger for each one to make itself the term or head of species and "decide" the time when each must live for itself from then on. As a jet of water breaks up into droplets towards the top of its stream, in the same way the phenomenon of "granulation of phyla" is revealed to our experience. (*Man's Place in Nature*, 1949, VIII, 93 E; 134 F)

GRAVITY *see also* **Weltstoff**

Movement of attraction in two inverse but complementary directions:

1 descent of Weltstoff towards increasingly probable states.
2 ascent of Weltstoff, using every opportunity, through greater complexity, towards increasingly improbable and increasingly conscious states.

There may well be a hidden relation between newtonian gravity of condensation (producing the stars) and "gravity" of complexification (producing life). In any case, the two can only function together. (*Man's Place in Nature*, 1949, VIII, 33 n.2 E; 47 n.2 F)

GROPING

Probing. Teilhard uses "groping" to express the idea of progression by trial and error. Multiplicity of attempts, efforts, in many different directions, preparing a breakthrough and forward leap. Teilhardian "keyword".

(a) And here is where the fundamental technique of groping continues, reappearing at the level of animate particles, that specific and invincible arm of every expanding multitude. Groping, where the blind fantasy of large numbers and the precise orientation of a pursued goal are combined. Not simply chance, as we often think, but directed chance. Filling everything to try everything, trying everything to find everything. (*The Human Phenomenon*, 1938–40, I, 66 E; 116 F)

(b) Without tentative gropings and without failures, without death, without planetary compression, as human beings, we would have remained stationary. (*The Activation of Human Energy*, 1953, VII, 305 E; 319 F)

GROWTH, AGGREGATES OF *see* **Growth, Aggregations of**

GROWTH, AGGREGATIONS OF

One of the three major factors making up the tree of life (the two others being the disjunctions (or breaches) that produce the verticils and the effects of distance that suppress the peduncles).

Aggregations of growth, giving birth to "phyla". (*The Human Phenomenon*, 1938–40, I, 69 E; 120 F)

GROWTH, PASSIVITIES OF

Dependence on energies that transcend the human but that are received and integrated into being in order to grow.

We answer to, and "communicate" with, the passivities of growth by our fidelity in action. Hence by our very desire to experience God passively we find ourselves brought back to the lovable duty of growth. (*The Divine Milieu*, 1926–7, 1932, IV, 58–9 E; 81 F)

HAPPINESS, EXISTENTIAL DIALECTICS OF

Three-stage dialectics of personalization:

- first stage = centration;
- second stage = decentration;
- third stage = excentration.

HEART, SACRED *see also* Christ, Heart of; Fire; Glow, Golden

Symbol of the divine. Center of a convergent universe. Focal point of evolution. For Teilhard, the Feast of the Sacred Heart is the feast of fire. He sees "fire capable of penetrating everything – and gradually spreading everywhere . . . " (*The Heart of Matter*, 1950, XIII, 47 E; 58 F).

(a) It would be difficult for me to make anyone understand how deeply, vigorously and constantly . . . my pre-war religious life developed under the sign and in wonder of the Heart of Jesus. (*The Heart of Matter*, 1950, XIII, 43–4 E; 54 F)

(b) The Heart of Christ at the heart of matter. The "Golden Glow" as I like to say in English. (Letter to Jeanne-Marie Mortier, 30 October 1948, *Lettres à Jeanne Mortier*, 48)

HELL

For catholics, the existence of hell is an article of faith but that some human beings are or will be damned is not – even if, given free will and evil, it is a logical possibility.

(a) You have told me, O God, to believe in hell. But you have forbidden me to hold with absolute certainty that a single person has been damned . . . I shall accept hell, on your word, as a structural element of the universe . . . (*The Divine Milieu*, 1926–7, 1932, IV, 141 E; 189 F)

(b) The existence of hell . . . is simply a negative way of affirming that, of physical and organic necessity, we can only find our happiness and our fulfillment by continuing in fidelity to the movement that carries us towards the term of our evolution . . . Hell, one can never repeat too often, is only known to us and only has meaning to the extent that it occupies in our minds the inverse place to heaven as the opposite pole to God. (*Introduction to the Christian Life*, 1944, X, 164–5 E; 192–3 F)

HOLINESS
Sanctity. State of religious perfection and spiritual purity attained, not by an ascesis seeking to suppress matter, but by an ascesis tending towards the spiritualization and sublimation of matter.

> The divine operation reigns in absolute sovereignty in the sphere of holiness . . . "Fold, my soul, your wings you had spread wide to soar to the terrestrial peaks where the light is most ardent. And wait the descent of the fire, if it wishes to take possession of you . . . True union that you should seek with the creatures that attract you, is found, not by going directly to them, but by converging with them towards God who is sought through them". (*The Mystical Milieu*, 1917, XII, 143 E; 185–6 F)

HOLISM *see* **Wholism**

HOMINIZATION *see also* **Anthropization; Humanization**
Appearance and development of the human phenomenon (principal phenomenon of noogenesis). Critical point. Passage from non-reflective animal life to reflective human life. Human potentialities are progressively realized in the world. Forces in the world are progressively spiritualized in human civilization. Teilhardian term for what Julian Huxley (1887–1975) calls "progressive psychosocial evolution".

> The human species progresses only in slowly elaborating from age to age the essence and totality of a universe deposited within us. To this great process of sublimation, it is fitting we apply the term hominization in its fullest sense . . . the individual instantaneous leap from instinct to thought, but that also in the wider sense is the progressive phyletic spiritualization in human civilization of all the forces contained in animality. (*The Human Phenomenon*, 1938–40, I, 122 E; 199 F)

HOMINIZATION, ELEMENTARY *see also* **Anthropization; Reflection, Step of**

> Elementary hominization or breakthrough to reflection. (*My Fundamental Vision*, No. 6, 1948, XI, 170 E; 187 F)

HOMINIZE *see also* **Hominization**

> In essence, and provided they maintain their vital connections with the current that rises from the depths of the past, are not the artificial, the moral and the juridical quite simply the natural, the physical and the organic, hominized? (*The Human Phenomenon*, 1938–40, I, 155 E; 246 F)

HOMO PROGRESSIVUS

> *Homo progressivus* . . . the human for whom the future of the earth is more important than the present. (*The Planetization of Mankind*, 1945, V, 137 E; 173 F)

HOMO SAPIENS SAPIENS

Only surviving species of the genus *Homo*.

> The appearance of *Homo sapiens* – not the end of hominization, but actually the beginning of true hominization or, one might simply say, a second hominization. (*The Phyletic Structure of the Human Group*, 1951, II, 150 E; 208–9 F)

HOMOGENEITY *see also* **Homogeneity, Principle of**

> The more we artificially cleave and pulverize matter, the more we can see its fundamental unity. In its most imperfect form, the simplest to imagine, this unity expresses itself in an astonishing similarity between the elements encountered . . . There is a unity of homogeneity, therefore. (*The Human Phenomenon*, 1938–40, I, 12–13 E; 35 F)

HOMOGENEITY, PRINCIPLE OF

General concept that gives every object of experience a certain permanence of dimension and axial structure of every object of experience through metamorphoses of growth to satisfy both change and identity.

> The first condition imposed on every object by our experience, if it is to be real, is not to remain always identical with itself or, on the contrary, to change constantly but to grow while retaining some of its own dimensions that make it continually homogeneous with itself. All around us life is born of life or "pre-life", freedom is born of freedom or "pre-freedom". Similarly, in the field of belief, faith is born of faith . . . In religious psychology, this is made necessary by the principle of homogeneity governing the synthesizing transformation of nature. (*How I Believe*, 1934, X, 98 E; 119 F)

HUMAN *see also* **Noosphere; Phenomenon, Human; Sense, Human; Ultra-human**

The human is situated between the cosmic and the christic. It is both the highest form of arrangement achieved by cosmogenesis (characterized by the reflective) and the emergence of an unprecedented event that constitutes a whole whose unity becomes more and more apparent to modern human beings who are waking to a real "human sense".

> Governing the whole of the universe as it now finally appears in my experience is a particular way of seeing the human. I say "human" and not "man" on purpose in order to stress, on the level of basic understanding, that what influences my vision of humanity is neither social concentration nor zoological species but the (quasi-physico-chemical) perception of a certain extreme state achieved in its thinking (what we might call its "uranium") by the stuff of the universe. (*The Stuff of the Universe*, 1953, VII, 376 E; 399 F)

HUMANITY *see also* **Human; Sense, Human**
Concrete and organic expression that is less notional and less abstract than the "human".

> As a collective reality, and therefore in a class of its own, humanity can be understood only insofar as we go beyond its body of tangible constructions and try to determine what particular type of conscious synthesis is emerging from its laborious and industrious concentration. This can ultimately be defined only in terms of spirit. (*The Human Phenomenon*, 1938–40, I, 175 E; 275 F)

HUMANIZATION *see also* **Hominization**
Increase in human characteristic due to human effort. By repercussion, penetration of the universe with human intentions. Humanization is cultural while hominization is biological.

> The nineteenth century and the early years of the twentieth were basically concerned with throwing light on the human past, the object of research being to establish clearly that the appearance of thought on earth corresponded biologically to a "hominization" of life. We now find that the present direction of scientific research, focussed ahead on the extensions of the "human phenomenon", is opening up an even more surprising perspective: that of a progressive humanization of humanity. (*Christ the Evolver*, 1942, X, 140 E; 165 F)

HYPER-
Prefix. Notion of integration, expanded scope, superiority beyond a given threshold, e.g. hyper-chemistry, hyper-concentration, hyper-einsteinian, hyper-orthodoxy, hyper-personal, hyper-physics, hyper-socialization, hyper-space, hyper-synthesis . . .

HYPER-EINSTEINIAN *see also* **Space–time**
Development of the einsteinian notion of space–time as a curve of convergence and arrangement.

> For every modern mind (to the extent that it is truly modern) consciousness is always apparent – a sense is born – in an absolutely specific universal movement by virtue of which the totality of things, from the highest to the lowest, is shifting together, as a whole, not only in space and time, but in a ("hyper-einsteinian") space–time whose particular curve makes everything that moves within it increasingly arranged. (*From Cosmos to Cosmogenesis*, 1951, VII, 256–7 E; 264–5 F)

HYPER-ORTHODOXY

> In classical theology, one might say, dogma was presented to our reason as a series of independent circles distributed on a plane. Today, reinforced by

a new dimension (the Universal Christ), the same pattern is tending to develop and group organically on a sphere in space – the simple and marvelous effect of hyper-orthodoxy. (*Suggestions for a New Theology*, 1945, X, 183 E; 213 F)

HYPER-PERSONAL

Super-concentration of persons at the term of cosmogenesis in God-Omega who, in the unity of the Three Persons of the Trinity, accomplishes in himself all the requirements of the personal while remaining universal and transcendent, beyond all anthropomorphic representation. Notion applied to:

- a collectivized humanity that transcends itself in omega and
- a God-Omega who constitutes the essential and transcendent hyper-personal. Consequently, the hyper-personal already exists in God but is still to be realized in humanity.

Nothing seems to me more feasible and fruitful (and consequently more imminent) than a synthesis between the above and the ahead in a becoming of a "christic" type in which access to the transcendent hyper-personal would appear conditional on the previous arrival of human consciousness at a critical point of collective reflection: the supernatural, consequently, would not exclude but rather require, as a necessary preparation, the complete maturation of the ultrahuman. (*Reflections on the Scientific Probability and the Religious Consequences of an Ultra-human*, 1951, VII, 279 E; 289–90 F)

HYPER-PHYSICS *see also* Ultra-physics

It is impossible to attempt a general scientific interpretation of the universe without seeming to intend to explain it right to the end. But only take a closer look at it and you will see that this "hyper-physics" still is not metaphysics. (*The Human Phenomenon*, 1938–40, I, 2 E; 22 F)

HYPER-SOCIALIZATION

Hyper-socialization . . . is nothing other than an ultra-vitalization (through ultra-arrangement) of the human mass that is gradually obliged to shift its position in a convergent universe. (*The Convergence of the Universe*, 1951, VII, 292 E; 305 F)

HYPER-SPACE

Four-dimensional space.

HYPER-SYNTHESIS

The time seems to have come when these (human) fragments will have to be combined and welded together under the irresistible pressure of geographical, biological, political and social determinisms accumulated on a planetary basis – this operation coinciding with the awakening of a genuine "spirit of the earth" that transcends the national spirits we have known so far. A new order of consciousness emerging from a new order of organized complexity. A hyper-synthesis of humanity upon itself. (*Universalization and Union*, 1942, VII, 89–90 E; 96 F)

I

IDENTIFICATION *see also* **Identification, Pantheism of; Union**
Opposite of union. Fusion and confusion of egos in an impersonal power (elementary matter, misunderstood collectivity, the impersonal God of the pantheisms).

> In eastern (or hindu) thought there is no doubt it is the ideal of diffusion and identification that has been dominant from the beginning. The elementary egos are seen as anomalies to be reduced (holes to be filled) in the universal being. *(Reflections on Two Converse Forms of Spirit*, 1950, VII, 224 E; 233 F)

IDENTIFICATION, PANTHEISM OF

> According to the first way (that I shall call more or less for convenience the "Road of the East"), spiritual unification is conceived as being effected by a return to a common "divine" basis subjacent to, and more real than, all sensibly perceived determinants of the universe. From this point of view, mystical unity appears and is obtained by the direct suppression of the multiple, that is, by relaxation of the cosmic effort of differentiation in and around ourselves. Pantheism of identification. Spirit of "détente". Unification with the sphere by co-extension of dissolution. (*My Fundamental Vision*, No. 33, 1948, XI, 200 E; 215 F)

IGNATIUS, ST. *see also* **Exercises, Spiritual; Francis, St.**
Iñigo López de Recalde (1491–1556), founder of the Society of Jesus (1534).

(a) I dream of a new St. Francis or a new St. Ignatius who would show us the new sort of christian life (both more involved in and more detached from the world) of which we have need. (Letter to Auguste Valensin SJ, 21 June 1920, *Lettres intimes*, 77)

(b) Oh how I would like to have met St. Ignatius or St. Francis whom our age needs so much. (Letter to Auguste Valensin SJ, 31 December 1926, *Lettres intimes*, 144)

IMMANENCE *see also* **Transience; Radial; Things, Inside of**

1 In a traditional sense, presence in the real of a principle both interior and superior to it. Opposed to transcendence.

> The present postulate of divine immanence is doubly valid:

- firstly, because, in fact, we experience the presence *ad intra* of a reality that is more absolute and more precious than ourselves . . .
- secondly, because God, if he is to be God, must achieve for us the *summum* of intimacy. (*The Soul of the World*, 1918, XII, 185 E; 253 F)

2 In a specifically teilhardian sense, presence of an interiority that grows as it rises through the different levels of the real.

> We cannot doubt that what we might call basic matter is certainly animated in its own way. Complete exteriority or total "transience", like absolute multiplicity, are synonyms of non-being. Atoms, electrons, elementary corpuscles, no matter what they may be (provided they are something outside ourselves), must have a rudimentary immanence, that is, a spark of spirit. (*My Universe*, 1924, IX, 46–7 E; 75 F)

IMPLOSION

Opposite of explosion. Phenomenon of contraction and centration corresponding to a greater interiorization that is sparked by the conjunction of an ascending flow of immanence and a descending flow of transcendence. Leads to an internal rupture and burst of spiritual fire and light.

> A double sense (and sentiment) of cosmic convergence and christic emergence that, each in its own way, took complete control of me . . . Brought together, the two spiritual ingredients constantly reacted on each other with an extraordinary brilliance: releasing, by their implosion, such an intense light that it transfigured (or even "transubstantiated") for me the very depths of the world. (*The Christic*, 1955, XIII, 82–3 E; 97 F)

IMPLOSIVE *see also* **Implosion**

> Forced together ever more closely by the progress of hominization and drawn together even more by a fundamental identity the two Omegas (one of experience and the other of faith) are, I repeat, clearly getting ready to react on each other in human consciousness and finally to *become synthesized*: the cosmic being on the point of magnifying fantastically the christic; and the christic being on the point (improbably) of amorizing (or energizing to the maximum) the entire cosmic. A truly inevitable and "implosive" meeting whose probable effect tomorrow will be to weld science and mysticism together in an enormous wave of liberated evolutionary power around a Christ finally identified two thousand years after Peter's witness as the ultimate summit (as the only possible God) of an evolution definitively recognized as being convergent. (*The God of Evolution*, 1953, X, 242–3 E; 291 F)

IMPURITY *see also* **Purity**

Materialization and dissolution of personal energies in the multiple that destroys spiritual unity. Expression with both ethical value and ontological meaning.

Impurity . . . brings about the physical dissolution and materialization of beings in their individual structure. In embracing the multiple in order to become lost within it, we break in some mysterious but nevertheless real way the fragile links of our own spirituality. (*The Struggle against the Multitude*, 1917, XII, 103 E; 140 F)

INCARNATION *see* **Creation**

INDIVIDUAL *see also* **Individuality; Personal**

(a) In a universe in process of centration . . . the individual and the collective continually reinforce and complete one another. The more individuals are associated appropriately with other individuals, the more, under the influence of synthesis, they penetrate deeper within themselves, become aware of themselves and consequently become personalized. (*The Atomism of Spirit*, 1941, VII, 51 E; 58 F)

(b) The necessity and importance of not confusing the two notions, partially independent, of the personal and the individual. What makes a center "individual" is its being distinct from the other centers that surround it. What makes it "personal" is its being profoundly itself . . . Understood in a restricted sense, as defining not the distinction but the separation between beings, individuality decreases with centrogenesis and ceases to exist (in omega) when personality reaches its maximum. (*Centrology*, 1944, VII, 117 n.9 E; 123 n.1 F)

INDIVIDUALITY *see also* **Individual; Personal**

The living monad to the extent that it constitutes a distinct and autonomous whole:

1 whose evolutionary function is to be the link responsible for transmitting through reproduction the innate characteristics of the phylum and other acquired characteristics and
2 whose task is to facilitate the emergence of the person.

Egoism, whether it be individual or racial, is justifiably intensified by the idea that in fidelity to life element rises to the extremes of uniqueness and incommunicability it possesses within itself. It thus feels right. Its only mistake, but a fatal one that leads it to go astray, is to confuse individuality with personality. In trying to distinguish itself as much as possible from others, the element individualizes itself; but in doing so it falls back and tries to drag the rest of the world backward toward plurality and into matter. In reality, the element diminishes itself and becomes lost. To become fully ourselves, it is in the opposite direction we must advance, in the direction of convergence with all the rest, towards the "other". The end of ourselves and the culmination of our originality is not in our individuality, but in our person; and according to the evolutionary structure of the world, the only way we can find our person is by uniting with one another. (*The Human Phenomenon*, 1938–40, I, 187 E; 292 F)

INDIVIDUALIZATION *see also* **Individuation**

> By this individualization of itself in the depths of itself, the living element is formed for the first time into a punctiform center, where all representations and all experiences join together and consolidate into a whole that is conscious of its organization. (*The Human Phenomenon*, 1938–40, I, 110 E; 181 F)

INDIVIDUATION

Isolation of each conscious center in a first phase of the formation of the noosphere where socialization begins with the first phase of expansive dispersion. Individuation causes a separation of individuals while individualization is a process of progressive interiorization within each individual. Not to be confused with jungian terminology.

> Previously . . . the human soul appeared more obviously attracted towards its own center than drawn by the hopes of the race. Then the apparent texture of life suddenly changed . . . Through the abuse of our freedom, unwise or vicious, what had only been a delicate stage in the synthesis of spirit suddenly became an acute crisis, an almost fatal crisis . . . The appearance of the immortal soul led to a serious crisis of individuation in the world, a counter-attack by the multiple, diminishing, suffering, culpable. (*The Struggle against the Multitude*, 1917, XII, 104–5 E; 141–3 F)

INFALLIBILITY *see* **Assumption, Dogma of the; Church, Infallibility of the**

INFINITY, THIRD *see also* **Complexity, Infinity of**

Teilhardian infinity of complexity that completes the two pascalian infinities – the infinitely large and the infinitely small. Part of the process of complexification and unification that begins with an indefinite multiple and ends in the infinity of the divine center. The third infinity is both a third dimension and the synthesis of the first two infinities because its temporal complexifying function is linked to a space that was previously considered static.

> (a) Spreading through a counter-current across entropy, we find a cosmic drift of matter towards increasingly centro-complexified states of arrangement (in the direction of – or interior to – a “third infinity”, an infinity of complexity, as real as the infinitesimal and the immense). (*My Phenomenological View*, 1954, XI, 212 E; 233 F)
>
> (b) This “eu-corpusculization” . . . develops “transversally” towards the very small and the very large, in a special form of infinity as real as those (the only ones usually considered) of the infinitesimal and the immense: the third infinity of organized complexity. (*The Singularities of the Human Species*, 1954, II, 212–14 E; 300–2 F)

INTEGRISM *see also* **Modernism**

Theological conservatism.

(a) There are those who wish to identify christian orthodoxy with "integrism", that is, with respect for the tiniest wheels of a little microcosm constructed centuries ago . . . Integralism or integrism, dogma-as-axis or dogma-as-framework . . . Integrism is simple and convenient, both for the faithful and the authorities. But it implicitly excludes from the Kingdom of God (or denies, on principle) the huge potentialities whirling around us in social and moral questions, in philosophy, science, etc. (Letter to Léontine Zanta, 7 May 1927, *Letters to Léontine Zanta*, 79 E; 86–7 F)

(b) Basically, even between integrists and us, there is basic agreement on a fundamental principal I would call "the law of the maximum Christ": the orthodox believer is one who attaches to Christ the maximum historical reality, grandeur and consistence. We shall see finally who will "win the cup" in this competition. (Letter to Jeanne-Marie Mortier, 4 September 1950, *Lettres à Jeanne Mortier*, 70)

INTER-

Prefix. Between, among. Expresses the notion of reciprocal influences implicit in teilhardian thought, e.g. inter-action, inter-activity, inter-centric, inter-combine, inter-complexity, inter-fecundation, inter-human, inter-individual, inter-influence, inter-liaison, inter-penetrability, inter-psychic, inter-reaction, inter-reflection, inter-sympathy . . .

INTER-CENTRIC *see also* **Centration; Centricity; Centro-complexity; Centrogenesis, Centrology; Decentration; Eu-centrism; Excentration; Super-centration**

Center-to-center relationships, above all, on the level of personal centers and, by extension, on the level of infra-personal centers.

(a) The problem obviously consists in finding the way of grouping ourselves together, not "tangentially" in the nexus of an extrinsic activity or function, but "radially", center to center, in such a way as to stimulate though synthesis deep within ourselves a directly centric progress of nature. In other words, what we must do is to love one another since, by definition, love is equally the name we give to "inter-centric" actions. (*The Atomism of Spirit*, 1941, VII, 47 E; 54 F)

(b) Love is the driving force of inter-centric relationships. Present (at least in rudimentary form) in all natural centers, living or pre-living, that make up the world, it also represents the deepest, the most direct, the most creative form of interaction that can be conceived between centers. In fact, it is the expression and the agent of universal synthesis. (*The Rise of the Other*, 1942, VII, 70–1 E; 77 F)

INTERIORIZATION *see also* **Centration; Consciousness; Psyche**
Growing importance of the center that accentuates complexification.

> If the universe appears to be sidereally in process of spatial expansion (from the infinitely small to the immense), then in the same way, and even more clearly, it appears to be physico-chemically in process of organic enfolding (from the very simple to the extremely complicated) – this particular enfolding of "complexity" being linked experimentally to a corresponding increase of interiority, that is, of psyche or consciousness. (*The Human Phenomenon*, 1938–40, I, 216–17 E; 334 F)

INTERIORIZE *see also* **Interiorization**

(a) Taken as a whole, in its temporal and spatial totality, life represents the term of a large-scale transformation in the course of which what we call "matter" (in the most comprehensive sense of the word) is inverted, folded in on itself, interiorized, the operation covering, as far as we are concerned, the entire history of the earth. (*The Phenomenon of Spirituality*, 1937, VI, 96 E; 121 F)

(b) A certain mass of "basic" (or "exteriorized") matter, that has been used in the course of the spiritualization of the universe, must be "interiorized" at the end of this operation if it is to be successful (as infallibly it must be). So much matter, so much spirit. (*The Phenomenon of Spirituality*, 1937, VI, 99–100 E; 125 F)

INVENTION
Function of life giving birth to organic tools. Represents, at the human level, the creation of artificial tools and, more generally, work by which human beings express themselves.

> If our "artificial" constructions are nothing but the legitimate sequel to our phylogenesis, then invention also, that revolutionary act from which the creations of our thoughts emerge one after the other, can legitimately be seen to prolong in reflective form of the obscure mechanism by which every new form has always germinated on the trunk of life. (*The Human Phenomenon*, 1938–40, I, 156 E; 248 F)

INVOLUTION *see also* **Arrangement; Counter-evolution; Enfolding**
Entropy. Opposite of evolution in two different senses:

1 drift of the world that is opposed to the ascent towards the improbable by a sort of counter-arrangement of cosmic energy for the benefit of the unorganized plural and the probable. Antonym of arrangement.

> A universe of primal "material" stuff is irremediably sterile and fixed, whereas a universe of "spiritual" stuff has all the elasticity necessary to lend itself to both evolution (life) and involution (entropy). (*The Spirit of the Earth*, 1931, VI, 23 E; 29 F)

2 rolling inwards on itself of the cosmic stuff with a view to a centration where the ascending movement of evolutionary creation and the descending movement of the incarnation collaborate in such a way that the result is a rolling inwards that synthesizes within itself the two directions of evolution and involution. Synonym of counter-evolution and entropy.

(a) *with specifically religious overtones*:
The layers of the world around us appear much richer and more penetrating in the perspectives of a christic type of creation (where a descending divine involution is combined with an ascending cosmic evolution). (*My Fundamental Vision*, No. 24, 1948, XI, 191 E; 205–6 F)

(b) *with no specifically religious overtones*:
The human soul is the definitively purified form of being towards which the work of conscious involution that, in fact, is what evolution is all about, must ontologically lead. (Journal, 31 December 1916, Journal, 176 see also Bruno de Solages OCarm, *Teilhard de Chardin*, 102 n.84)

IRENÆUS, ST. *see also* **Fathers, Greek**

Irenæus (*c.*130–*c.*200), theologian and bishop. He sees the Cross, not only as the central event in human redemption, but also as the center of the whole universe.

One finds an astonishing anticipation of our modern views on progress in the Greek Fathers – Irenæus, for example. (*The Mysticism of Science*, 1939, VI, 167 n.1 E; 208 n.1 F)

IRREVERSIBILITY *see also* **Centro-complexity; Orthogenesis**

1 In a biological sense, property of a temporally-oriented process and, as such, fundamentally opposed to the pure and simple return to an earlier state. More general notion than orthogenesis.

It seems many of the difficulties encountered in the palaeontological application of the law of irreversibility arise from a confusion between irreversibility and orthogenesis. Clearly, the two notions are very different. Irreversibility is far from always pursuing a development in the same direction (orthogenesis). On the contrary, it allows, in the history of the forms that obey it, all sorts of counter-currents and circuitousness. (*The Law of Irreversibility in Evolution*, 1923, III, 49 E; 73 F)

2 In a phenomenological sense, positive property of the real to continue the development towards the ahead of more spiritual, more stable syntheses that are increasingly distanced from the danger of reversal, failure and, finally, death.

Irreversibility of the stream of life is proved, to a certain degree, by its very success; why should it go back if, on the whole, it has never ceased to grow from the start? We could add (and this proof is very strong if we can understand it) that life become reflective in the human seems to require, for its very functioning, that it shall be irreversible. (*The Phenomenon of Man*, 1930, III, 170 n.1 E; 129 n.2 F)

3 In a metaphysical sense, synonymous with immortality since, beyond a given threshold of centro-complexity, what has been made cannot be unmade.

Immortality or, in the very general sense in which I use the word, irreversibility, seems to me to follow, as a necessary property or complement, all idea of a universal progress. (*How I Believe*, 1934, X, 109 E; 129 F)

J

JESUS, SACRED HEART OF *see also* **Heart, Sacred**

JURIDICISM *see also* **Extrinsicism**

Opposite of organicism. Reduces morality to a legal code and the relationship between Christ and the world to the level of a monarch administering his land and people. Juridicism is linked to extrinsicism.

> Finally, christianity is specifically personalist. But, here again, has not the dominant role conceded to the values of the soul made christianity inclined to show itself, above all, as a form of juridicism and morality, instead of showing us the organic and cosmic splendors of the Universal Christ? (*The Salvation of Mankind*, 1936, IX, 150 E; 191 F)

JURIDICIST *see also* **Juridicism; Juridics**

> The pluralists always reason as if no principle or connection existed, or tended to exist, in nature outside the vague or superficial relations habitually examined by "common sense" and sociology. At heart, they are juridicists and fixists who cannot imagine anything around them, except what seems to them to have always been there. (*Human Energy*, 1937, VI, 146–7 E; 182 F)

JURIDICS *see also* **Juridicism; Juridicist**

> There are two categories of irreconcilable mind: the physicists (the "mystics") and the juridics. For the former, being is not beautiful unless it is organically structured; hence Christ, sovereignly attractive, must radiate physically. For the latter, being is a matter of concern when it conceals within itself something greater and less definable than human social relations (considered as being artificial). Accordingly, Christ is no more than a king and a landowner. (*My Universe*, 1924, IX, 55 E; 83–4 F)

L

LAMINATIONS, STRUCTURE IN *see also* **Structure, Scaly**

Though, in certain respects, humans stands apart from the anthropomorphs that most clearly resemble them, they are zoologically inseparable from the history of their group. If we still doubted the connection, it would be enough to observe how perfectly the structure in roughly concentric laminations, that is that of the entire order of primates, continues on a smaller but identical scale within the hominian group. (*Paleontology and the Appearance of Man*, 1923, II, 50 E; 73 F)

LAYER *see also* **Biota**

Evolutionary unit of living creatures that develops from a gradually disappearing peduncle (stalk). Represents a reality similar to (but different from) the biota.

Below . . . the history of the mammals is lost in darkness. All around and toward the top, at least, their group, naturally isolated by the rupture of its peduncle, stands out with enough clarity and individuality to allow us to take it as a practical unit of "evolutionary mass". Let us call this unit a layer. (*The Human Phenomenon*, 1938–40, I, 80 E; 137 F)

LEAF *see also* **Laminations, Structure in; Scale, Zoological**

Once we have identified the pithecanthropian leaf as such . . . we are encouraged to look in other places for traces of other similar units. (*Man's Place in Nature*, 1949, VIII, 68 E; 98 F)

LEVEL *see also* **Sphere**

If any single perspective has been clearly brought out by the latest advances in physics, surely it is that in the unity of nature there are different types of spheres (or levels) for our experience, each one characterized by the dominance of certain factors that become imperceptible or insignificant in the neighboring sphere or level. (*The Human Phenomenon*, 1938–40, I, 23 E; 50 F)

LIBERATION, THEOLOGY OF *see also* **Theogenesis**

In a neo-teilhardian sense, theology of faith and hope in the future of the human person on earth, a future built on an evolutionary consciousness that is ever reaching consummation in the presence of God, a radical eschatology. Teilhard looks to the liberation, not only of the noosphere (the sphere of human activity), but also of the biosphere and the geosphere (John Morgan, *Teilhard Review*, The British Teilhard Association, XI.1, 20; see also Ursula King, *Towards a New Mysticism*, 228).

(a) Love God in and through the universe in evolution: it is impossible to imagine a more constructive, more complete, more appealing, more precise formula for action that, however, is more open to the unforeseen demands of the future. As I said, a theoretical formula for action. But even more an actually living neo-mysticism in which there is an irresistible tendency to combine in every modern mind under the christian banner the two basic attractions that hitherto split the human power to worship between heaven and earth, between theo- and anthropocentrism. (*Christianity and Evolution*, 1945, X, 185 E; 216 F)

(b) The future of the thinking earth is organically linked to the transformation of the forces of hatred into forces of charity. (*Natural Human Units*, 1939, III, 214 E; 300 F)

LIFE *see also* **Phyletic**
Threshold of discontinuity in the complexification of matter and, consequently, in the development of consciousness. Movement from the molecular (pre-centric) to the phyletic. Second of the four sections (pre-life, life, thought, superlife) in *The Human Phenomenon*.

> Life is nothing other, scientifically speaking, than a specific effect (the specific effect) of complexified matter. (*Man's Place in Nature*, 1949, VIII, 24 E; 34 F)

LIFE, BREAKTHROUGH TO *see also* **Life, Step of**

LIFE, FULLER *see also* **Being, Better**
Movement of transcendence at the center of life when the latter crosses the higher threshold of reflective consciousness and, consequently, finds itself controlled, directed and integrated by spiritual energies. Represents the final phases of both the transphenomenal threshold and the spatio-temporal.

> An open world? Or a closed world? A world ultimately emerging into some fuller life? Or a world finally falling back with its whole weight? (*The Activation of Human Energy*, 1953, VII, 391 E; 414 F)

LIFE, STEP OF
Passage from the pre-living molecule to the living mega-molecule, capable of auto-construction and reproduction. Marks a discontinuous emergence in a process of continuous centration.

> The external realization of an essentially new type of corpuscular grouping

that permits the more supple and better-centered organization of an unlimited number of substances taken at all degrees of particulate size; and simultaneously the internal appearance of a new type of conscious activity and determination: it is by this double and radical metamorphosis that we can reasonably define what is specifically original in the critical passage from molecule to cell – the step of life. (*The Human Phenomenon*, 1938–40, I, 50–1 E; 92 F)

LOGICALISM *see also* **Fixism; Physicism**

Anti-evolutionary concept (special case of a general concept of the real) according to which living creatures are distributed in the universe according to a purely intellectual and logical plan. Concept excludes the chronological and genetic point of view.

A first, and theoretically possible, answer to the question (on the nature of the "something" that binds together the elements that constitute living creatures) is: "Living creatures are distributed in the universe according to a purely intellectual plan. Between these various forms there is no bridge of determinism, no connection of a physical nature, only an artificial continuity. The law of succession of living creatures, the reason for their similarity, is not to be found inside things. It is wholly concentrated in a creative idea that develops, at successive points, in a given series, the design it has conceived in its wisdom . . . Living forms constitute a chain, they lead on from one to another, thanks to a logical sequence in the mind of God". This theory can be called logicalism. (*Note on the Essence of Transformism*, 1920, XIII, 109–10 E; 126–7 F)

LOVE *see also* **Amorization**

Unitive and differentiating energy originally flowing from the divine focal point, having emerged, par excellence, in the charity of Christ. Activates person-to-person and center-to-center relationships. Not to be confused with purely sentimental affection. Love is the very sap of creative union and the mark, in the person, of the convergence of the universe. Love is the fundamental cosmic energy that moves the universe. Teilhard sings with Dante of the "love that moves the sun and the other stars" (*The Divine Comedy, Paradise*, 1962, 347).

(a) Love is the most universal, the most tremendous and the most mysterious of cosmic forces . . . Love is the primal and universal psychic energy . . . Love is a sacred reserve of energy . . . the blood of spiritual evolution. (*The Spirit of the Earth*, 1931, VI, 32–4 E; 40–2 F)

(b) Love is nothing more or less than the direct or indirect trace marked in the heart of the element by the psychic convergence of the universe upon itself. (*The Human Phenomenon*, 1938–40, I, 188 E; 294 F)

(c) A love that embraces the entire universe is not only something psychologically possible; it is the only complete and final way in which we can love. (*The Human Phenomenon*, 1938–40, I, 190 E; 296 F)

(d) Love is the driving force of inter-centric relationships. Present (at least in rudimentary form) in all natural centers, living or pre-living, that make up the world, it also represents the deepest, the most direct, the most creative form of interaction that can be conceived between centers. In fact, it is the expression and the agent of universal synthesis. (*The Rise of the Other*, 1942, VII, 70–1 E; 77 F)

(e) Love not only has the virtue of uniting without depersonalizing but also of ultra-personalizing by uniting. (*Human Energy*, 1937, VI, 154 E; 190 F)

(f) The problem obviously consists in finding the way of grouping ourselves together, not "tangentially" in the nexus of an extrinsic activity or function, but "radially", center to center, in such a way as to stimulate though synthesis deep within ourselves a directly centric progress of nature. In other words, what we must do is to love one another since, by definition, love is equally the name we give to "inter-centric" actions. (*The Atomism of Spirit*, 1941, VII, 47 E; 54 F)

(g) What name should we give . . . to the physico-moral energy of personalization to which every activity revealed by the stuff of the universe is finally reduced? Only one – love: if we are to credit it with the generality and power it must assume on rising to the cosmic order. (*Sketch of a Personalistic Universe*, 1936, VI, 72 E; 90F)

LOVE, PANTHEISM OF *see also* Pantheism; Union, Pantheism of

According to the second way (the Road of the West) . . . it is impossible to become one with all without pushing to the limit, in their simultaneous direction of differentiation and convergence, the dispersed elements that constitute and surround us. From this second point of view, the "common basis" of the eastern way is no more than an illusion: only a central focal point exists at which we can only arrive by pushing to their junction the countless guidelines of the universe. Pantheism of union (and hence of love). Spirit of "tension". Unification by concentration and hyper-concentration at the center of the sphere. (*My Fundamental Vision*, No. 33, 1948, XI, 200–1 E; 215–16 F)

MARXISM *see also* **Terrenism**

Synonymous in teilhardian thought with terrenism. Not intended as a definitive or historical definition of marxism. Teilhard is less attracted to marxist dialectical materialism than its concrete and human aspects to which he seeks to give a religious dimension. No spiritual concession is ever made to marxism.

(a) In fact, every conversation I have ever had with communist intellectuals has left me with the definite impression that marxist atheism, far from being absolute, merely rejects the "extrinsical" sort of God, a "Deus ex machina", whose existence can only harm the dignity of the universe and weaken the source of human effort: a "pseudo God", in short, no one today (and not least the christians) wants any longer. (*The Heart of the Problem*, 1949, V, 266–7 E; 346 F)

(b) The humanist pantheisms around us represent a wholly youthful form of religion. A religion little or even uncodified (outside marxism). A religion without an apparent God, without any revelation. But a religion in a real sense, if we understand by this word a contagious faith to which you can give your life. Despite great differences of detail, a rapidly growing number of our contemporaries already agrees on recognizing the supreme value of existence in devoting oneself body and soul to a universal progress that expresses itself through the tangible developments of humanity. (*How I Believe*, 1934, X, 123 E; 143 F)

MASS, THE *see also* **Eucharist; Eucharistization**

Eucharist. One of the seven sacraments recognized by the Catholic and Orthodox Churches.

(a) From age to age there is but one Mass: it is the universe that is the true host, the total host, a universe that is ever more intimately penetrated and vivified by Christ . . . Fundamentally, from the beginning and for ever, there is but one thing that is being made in creation: the Body of Christ. (*Pantheism and Christianity*, 1923, X, 73–4 E; 90 F)

(b) All the communions of a lifetime form but one communion. All the communions of all men and women now living form but a single communion. All the communions of all men and women, past, present and future, form but a single communion. (*The Divine Milieu*, 1926–7, 1932, IV, 113 E; 151 F)

MATERIA MATRIX *see also* **Matter, Spiritual Power of**
The material matrix of spirit. Metaphor that suggests:

1 matter is essential to the genesis of spirit;
2 spirit tends to become detached from its material infrastructure in order to come within the zone of attraction of omega.

(a) Like the child still within its mother our spirit depends on the fibers and roots that extend into materia matrix. It needs this to live. Its role is to extract, to the last drop if possible, the spiritual power largely spread around the lower circles of the universe. (*The Names of Matter*, 1919, XIII, 231 E; XII, 456 F)

(b) Matter is the matrix of spirit. Spirit is the higher state of matter . . . Matter is the matrix of consciousness and all around us matter, born of consciousness, is constantly advancing towards some ultra-human. (*The Heart of Matter*, 1950, XIII, 35, 45 E; 45, 56 F)

MATERIALISM
Expression that seeks to explain all things by their (constituent) elements. Consequently, the consistence of the world depends on the elementary on which the cosmos will finally settle in equilibrium. The teilhardian definition responds to classical materialism rather than dialectical materialism.

The first idea that comes to upstart humans, through scientific analysis, to the very lowest limits of matter, is that we really hold in the ultimate particles of matter the very essence of the riches of the universe. "The elements contain within themselves the quality of the whole: whatever holds the elements possesses the whole". This principle is implicitly admitted by many scientists and not a few philosophers . . . If this principle were true, it would be necessary to say we are driven by science to materialism. (*Science and Christ*, 1921, IX, 27 E; 53 F)

MATERIALIZATION *see also* **Matter; Spirit-Matter**
Synonym of matter within the framework of a dynamic, genetic perspective where matter and spirit are presented as simply conjugated variants.

(a) There is neither spirit nor matter, there is one
| spiritualization
| materialization. (Journal, 24 December 1920, in Bruno de Solages OCarm, *Teilhard de Chardin*, 286 n.59)

(b) Within what is no longer a cosmos but a cosmogenesis, through the successive thresholds of materialization, vitalization and reflection, one and the same energy, one and the same solidarity, is built up. (*The Evolution of Responsibility*, 1950, VII, 213 E; 220 F)

MATTER *see also* **Spirit-Matter**

Matter is multiple. Spirit is one.

1 In a conceptual sense, pure matter is the antithetical multiple of the one. Pure matter is both opposed to, and the object of, unitive energy. As multiple, pure matter is the subject of analysis, while unified being is the product of synthesis.

2 In an evolutionary (phenomenal) dimension, there is neither pure matter nor pure spirit but only spirit-matter that is either in process of progressive spiritualization (through unification) or in danger of regressive materialization (through falling back into the multiple). Process characterized by numerous names of matter (see below).

(a) Basically, matter in a being (in a monad) is what makes the being capable of being united to other beings, in such a way as to form with them a newer, simpler whole. Matter is not what unites (only spirit unites). But it lays itself open to union. (*The Names of Matter*, 1919, XIII, 227 E; XII, 450–1 F)

(b) Matter, from the ascetic or mystical point of view adopted in these pages, does not exactly fit any of the abstract entities defined by this name by science or philosophy. It is certainly the same concrete reality for us as for physics or metaphysics, with the same basic attributes of plurality, of tangibility, of interconnections. But we are trying to embrace this reality as a whole, in the widest possible sense: we take account of its full abundance as it reacts, not only to our scientific or dialectic inquiries, but to all our practical activities. Matter, as far as we are concerned, will be the ensemble of things, of energies, of the creatures about us, to the extent that these are palpable, sensible (feeling), "natural" (in the theological sense of the word). This will be the common, universal, tangible milieu, infinitely varied and shifting, in which we live. (*The Divine Milieu*, 1926–7, 1932, IV, 89–90 E; 121–2 F)

(c) Matter is the matrix of spirit. Spirit is the higher state of matter . . . Matter is the matrix of consciousness and all around us matter, born of consciousness, is constantly advancing towards some ultra-human. (*The Heart of Matter*, 1950, XIII, 35, 45 E; 45, 56 F)

I THE NAMES OF MATTER

(1) Formal matter. Matter considered as the principle of action. The scholastic *subjectum* in a concrete dimension. (*The Names of Matter*, 1919, XIII, 227 E; XII, 450–1 F)

(2) Concrete matter. Matter considered as existing (as opposed to the formal principle). Defined by reference to a process of union ahead and an abyss of dispersion behind. (*The Names of Matter*, 1919, XIII, 227–8 E; XII, 451–2 F)

(3) Universal matter. Matter considered at an inchoate, rudimentary stage. Unity of the whole in still indistinct form. (*The Names of Matter*, 1919, XIII, 228 E; XII, 452 F)

(4) Total matter (see also II.2 below). Matter considered at a later, more

developed stage where its connections converge in the spirit. (*The Names of Matter*, 1919, XIII, 228–9 E; XII, 452–3 F)

(5) Relative matter. Matter considered at all stages where:

(a) it represents a spiritual principle of union for the multiple that is inferior to it;
(b) it constitutes itself a multiple capable of being synthesized in a principle of higher union.

Comprises on the human level three applications that can be thought of as a ternary union:

A. Unifiable living matter. Spiritualizable matter. Matter destined to lose its materiality (antonym of inverse or dead matter). (*The Names of Matter*, 1919, XIII, 230–1 E; XII, 455–6 F)

B. Inverse or dead matter. Part of matter that is incapable of spiritualization. Returns to non-being. (*The Names of Matter*, 1919, XIII, 231–3 E; XII, 456–8 F)

C. Secondary or new matter. Appears after living matter is detached from inverse matter. Results from the condition inherent in all purified matter, from the consolidation and mechanization of results. Makes possible and brings about a new process of purification, both similar to and developed from the earlier process. (*The Names of Matter*, 1919, XIII, 233–4 E; XII, 458–60 F)

(6) Liberated matter. Transfigured matter that survives the metamorphosis of death and is offered at the resurrection. (*The Names of Matter*, 1919, XIII, 234–6 E; XII, 460–2 F)

(7) Resuscitated matter. Final phase of matter. Result of:

(a) excessive metamorphoses of matter before and in death and
(b) creative action, grace and divine efficiency. (*The Names of Matter*, 1919, XIII, 236–8 E; XII, 462–4 F)

II THE HUMAN PHENOMENON

(1) Elemental matter. Product of analysis of any artificially detached fragment of matter, analysis that brings out three basic properties:

(a) plurality;
(b) unity;
(c) energy. (*The Human Phenomenon*, 1938–40, I, 12–14 E; 34–7 F)

(2) Total matter (see also I.4 above). More developed notion. Matter considered in its concrete, spatio-temporal reality, as an evolutionary synthesis under three complementary heads:

(a) the system that controls plurality;
(b) the totum that reveals the design of the whole;

(c) the quantum that measures the global capacity of cosmic energy. (*The Human Phenomenon*, 1938–40, I, 14–16 E; 37–41 F)

MATTER, JUVENILE *see also* **Life; Pre-life**

Original substance of the earth. Contains evidence of pre-life and life.

> Geologists are still uncertain about the way the individualization of the earth took place – an agglomeration, in any case, of elementary particles. The basic idea to keep in mind at the moment is the extraordinary richness and complexity of its "juvenile matter", magma where, side by side with physico-chemical activities, that have become either neutralized or evaporated today, we find, in a form inaccessible to present knowledge, the influence of pre-life. (*The Spirit of the Earth*, 1931, VI, 25 E; 32 F)

MATTER, ORIGINAL *see also* **Matter-matter**

Matter-matter in its original position in evolution.

> Taken at its lowest point . . . original matter is something more than the particulate swarming so marvelously analyzed by modern physics. Beneath this initial mechanical sheet we must think of a "biological" sheet, thin in the extreme, but absolutely necessary to explain the state of the cosmos in the times that follow. (*The Human Phenomenon*, 1938–40, I, 24–5 E; 53 F)

MATTER, PRIMARY

Matter in its initial state of multiplicity on which the force of union and centration has not yet begun to operate.

> Repulsion in face of an inevitable totalization that threatens us with being sucked into a sort of "secondary matter" made up of accumulated determinisms. Fear in face of an end through mechanization, an end as dreaded as death through disintegration and return to "primary matter". (*The Heart of Matter*, 1950, XIII, 50 E; 61 F)

MATTER, SECONDARY *see also* **Matter**

> Repulsion in face of an inevitable totalization that threatens us with being sucked into a sort of "secondary matter" made up of accumulated determinisms. (*The Heart of Matter*, 1950, XIII, 50 E; 61 F)

MATTER, SPIRITUAL POWER OF

Capacity of matter to be penetrated and progressively transformed by unitive energy (whose presence is one of the essential effects of creation). Facilitates a universal genesis of spirit through matter and provides a means of sanctification within the perspectives of an ascesis by traversing and transcending. Matter is the starting point of a process of spiritual ascent that finds its completion in God.

> The idea of a universal genesis of spirit through matter (the idea, in other

words, of a spiritual power of matter) goes beyond, in its origins, the problem of chastity. (*The Evolution of Chastity*, 1934, XI, 69 E; 75 F)

MATTER-MATTER *see also* **Matter, Primitive**

Ideal limit of the multiple, symbolized by the most rudimentary forms of the mineral kingdom.

First of all, of course, forming the solid permanent core of the system is the taste for geology. The primacy of matter-matter is expressed in the mineral and the rock. (*The Heart of Matter*, 1950, XIII, 21 E; 29 F)

MAXIMUM, PRINCIPLE OF THE

Law that expresses the need for ontological balance between the structure of the real as a totality and the requirements of rationality and love nourished by the human monads that are part of this totality.

"By organic and metaphysical necessity, the world cannot be inferior, in coherence or value, to the ultimate demands of our reason and hearts". Or, in a positive sense, "What our reason and our hearts basically and positively require in order to be satisfied is what the world possesses". Or, again, "The most intelligible and the most activating is necessarily the most real and the most true". (*Action and Activation*, 1945, IX, 175 E; 221–2 F)

MEGA-

Prefix. Usually expresses the notion of size, grandeur, that, when placed before a unit, multiplies it by a million, but sometimes used to give intensity to composite words, e.g. mega-cephalic, mega-molecule, megasynthesis . . .

MEGA-MOLECULE

Directed complication is the law in which, from the micromolecules, then megamolecules, the very process ripened that gave birth to the first cells, and to it biology has given the name of orthogenesis. Orthogenesis is the only complete and dynamic form of heredity. (*The Human Phenomenon*, 1938–40, I, 65 E; 114–15 F)

MEGASYNTHESIS

I positively see no other coherent, and therefore scientific, way of grouping this immense succession of facts, except by interpreting the "super-arrangement" that all thinking elements of the earth find themselves subject to today, individually and collectively, in the sense of a gigantic psycho-biological operation as a kind of megasynthesis. Megasynthesis in the tangential. And therefore, by this very fact, a leap forward of radial energies along the principal axis of evolution. Still more complexity: and therefore even more consciousness. (*The Human Phenomenon*, 1938–40, I, 172 E; 271 F)

META-

Prefix. Expresses the notion of succession or change, e.g. meta-biology, metamorphosis, meta-phenomenon, metaphysics, meta-psyche . . .

METAMORPHOSIS, PAIN OF *see also* **Personalization, Pain of; Plurality, Pain of**

Positive effort and suffering necessary to cross the threshold between an old equilibrium that no longer exists and a new equilibrium that does not yet exist.

> If the pain of differentiation, inherent in union, generally has little effect on us, it is because we palpably associate it with consciousness of our progress. Much more bitter is the anguish of feeling ourselves apparently threatened by what lies most deeply hidden in our hearts. We can truly say that real suffering entered the world with the human when, for the first time, a reflective consciousness became capable of observing its own diminution. The only real evil is the "personal evil". How does death in the personal universe we have outlined here present itself? I would reply: "As a metamorphosis". (*Sketch of a Personalistic Universe*, 1936, VI, 87 E; 108 F)

META-PHENOMENON

That which is beyond the phenomenal, that is, beyond the spatio-temporal.

> Spirit is neither super-imposed on nor accessory to the cosmos . . . Spirit is neither a meta- nor an epi-phenomenon: it is the phenomenon. (*The Phenomenon of Spirituality*, 1937, VI, 94 E; 118 F)

METAPHYSICS *see also* **Dialectics; Morality; Mysticism; Recurrence, Law of phenomenological; Ultra-physics**

1 Deductive explanation of the world from abstract, absolute, non-temporal principles that constitute a definitive system.

> I should be happy to see you . . . substituting for a metaphysics that is stifling us an ultra-physics in which matter and spirit would be englobed in one and the same coherent and homogeneous explanation of the world . . . Thought will explode or evaporate unless the universe, in response to hominization, becomes divinized in some way . . . Christianity is the only living phylum that retains a divine personality. (Letter to Christophe Gaudefroy, 11 October 1936, *Lettres inédites*, 110, 111–12)

2 Second or third stage of the teilhardian vision (that, depending on the writing in question, consists of three or four stages). Continues the law of phenomenological recurrence while constituting a discontinuous and specific ontological threshold. Introduces:

(a) a new method: the systematic deduction from undoubtedly absolute principles but only producing a provisional, non-definitive synthesis;

(b) a new vision of being whose basic motivation is creative union;
(c) philosophical–theological developments of the triune God, creation, incarnation (problem of evil) and Pleroma.

In this second part, starting from certain general principles considered as being absolute, I shall try to reconstruct deductively (that is, a priori) the system described (including its theological or revealed (revelational) developments) . . . In classical metaphysics it is usual to deduce the world from the notion of being, considered irreducibly original. Backed by recent research in physics that has proved (contrary to "popular" belief underlying the whole of philosophia perennis) that movement is not independent of the moving body, rather that the moving body is physically engendered (or, more exactly, co-engendered) by the movement that inspires it, I shall try to show that a dialectic richer and suppler than others is possible, if from the outset, being, far from representing a terminal and solitary notion, is seen in reality as definable (at least genetically, if not ontologically) by a particular movement indissolubly associated with it – metaphysics of union. (*My Fundamental Vision*, No. 26, 1948, XI, 192–3 E; 207–8 F)

3 Any total vision of the real, demonstrated or postulated. Basis of modern metaphysics. Sense rarely used by Teilhard.

"Metaphysics" should be understood as meaning any solution or vision of the world (of life) "as a whole" (any Weltanschauung) that is either imposed on the intelligence or categorically adhered to as an option or postulate . . . Morality and metaphysics are inevitably to be seen (structurally) as the two faces (intellectual and practical) of one and the same system. Metaphysics is necessarily backed by morality and vice versa. Every metaphysics entails its own morality and every morality implies its own metaphysics. (*Can Moral Science dispense with a Metaphysical Foundation?*, 1945, XI, 130, 131 E; 143, 144 F)

META-PSYCHE

In conformity with the latest views that point us towards the idea of a spiritual essence of matter, will not physics end up by isolating and controlling whatever is hidden in the depths of the meta-psyche? (*Human Energy*, 1937, VI, 130 E; 163 F)

MILIEU *see also* Center; Circle; Milieu, Divine; Milieu, Mystical; Sphere

Fr. *mi-* (Lat. *medius*, mid) + *lieu* (Lat. *locus*, place), center, circle, heart, place, sphere. Synonym for center, environment. Teilhard uses "milieu" in its French sense to express both center and circle (or sphere). Hence the "divine milieu" is both the divine center and the divine circle, the divine heart and the divine sphere.

(a) The divine milieu is, in reality, a center . . . the ultra-active point of the universe. (*The Divine Milieu*, 1926–7, 1932, IV, 102, 103 E; 137, 138 F).
(b) God only reveals himself everywhere, beneath our gropings, as a universal

milieu because he is the ultimate point upon which all reality converges . . . It is precisely because God is the center that he fills the whole sphere. (*The Divine Milieu*, 1926–7, 1932, IV, 101, 102 E; 136 F)

MILIEU, DIVINE *see also* **Milieu; Milieu, Mystical**

Evolutionary super-milieu. Divine center and sphere. Field of divine energy that emanates from a focal point that centers, animates and directs it in its totality. The idea of milieu in a biological sense inspired Teilhard to transform it into the notion of a central focal point that is wholly immanent and wholly transcendent.

(a) With the universe christified (or, that comes to the same thing, with Christ universalized) an evolutionary super-milieu appears – that I have called "The Divine Milieu". (*The Christic*, 1955, XIII, 95 E; 110 F)

(b) Basically . . . what characterizes the divine milieu is its constituting a dynamic reality in which all opposition between universal and personal is going to be wiped out (without any confusion). The multiple "reflected" elements of the world will be completed in their infinitesimal ego by an integrated accession to the christic ego towards which the totality of participated being gravitates (and in consummating is consummated). (*The Christic*, 1955, XIII, 95 E; 110 F)

MILIEU, MYSTICAL *see also* **Milieu; Milieu, Divine**

First name given by Teilhard to the divine milieu with something of a hint of the personal election expressed in *The Mystical Milieu* (1917).

The mystical milieu does not constitute a completed zone in which beings remain immobile – once they have been able to become part of it. It is a complex element, composed of divinized creative being in which the immortal substance of the universe is gradually gathered in the course of time. We cannot call it God, rather his kingdom. We cannot say it is, rather it is becoming. (*The Mystical Milieu*, 1917, XII, 137 E; 179 F)

MILIEU, UNIVERSAL *see* **Milieu**

MODERNISM *see also* **Concordism; Integrism**

Began as an attempt to reconcile traditional catholic teaching with the new outlook in science and philosophy. Its critics saw it diminishing the person of Christ. Modernism was condemned by Pius X (decree *Lamentabili* and encyclical *Pascendi*, 1907). Priests were subsequently required to subscribe to an anti-modernist oath (motu proprio *Sacrorum Antistitum*, 1909)

(a) The modernist "volatilizes" Christ and dissolves him in the world. While I am trying to concentrate the world in Christ. (Cahier 7, 9 June 1919, in Bruno de Solages OCarm, *Teilhard de Chardin*, 342)

(b) In universalizing himself, Christ is not lost (as happened in the condemned forms of modernism) in the midst of the universe; but he dominates and assimilates the universe by imposing on it the three essential characteristics of his traditional truth: personal nature of the divine; manifestation of the supreme personality in the Christ of history; supraterrestrial nature of the world consummated in God. (*Some Reflections on the Conversion of the World*, 1936, IX, 123–4 E; 163 F)

(c) Basically, even between integrists and us, there is basic agreement on a fundamental principal I would call "the law of the maximum Christ": the orthodox believer is one who attaches to Christ the maximum historical reality, grandeur and consistence. We shall see finally who will "win the cup" in this competition. (Letter to Jeanne-Marie Mortier, 4 September 1950, *Lettres à Jeanne Mortier*, 70)

MOLECULIZATION *see also* **Corpusculization**

Chief phenomenon of biogenesis.

Hominization – the particular and (provisionally) definitive term of universal moleculization . . . Before the appearance of the human we could say nature was working to produce "the unity or grain of thought". Now it definitely appears, according to the laws of some gigantic hyper-chemistry, we are now being driven towards "edifices of grains of thought", towards a "thought of thoughts" – ever deeper in the abyss of the infinitely complex. (*The Atomism of Spirit*, 1941, VII, 34, 38 E; 41, 45 F)

MONAD *see also* **Monad, Great**

Physical or psychic element. Expression used to represent:

1 human individuality as an element of the whole.

To look at the universe, as the human phenomenon compels us to do, as composed of psychic nuclei, with each one playing the role of a partial center in relation to the world, being virtually co-extensive with the universe, is clearly to go back to the methods of Leibnitz. While cosmic corpuscles in the static universe of monadology "have neither doors nor windows", from the evolutionary point of view of centrology they are triply interdependent within the centrogenesis into which they are born. (*Centrology*, 1944, VII, 104 E; 109 F)

2 the unanimity of human monads constituting the individualized whole that, in turn, constitutes the Great Monad.

Strictly speaking, there is only one individuality (only one monad) in the universe, that of the whole (considered in its organized plurality). The unity or measure of the world is the world itself. (*The Universal Element*, 1919, XII, 296–7 E; 439 F)

3 the supreme center of personalizing energy.

Within the cosmos all the elements, in the ascending order of their true being (that is, of their consciousness), hang together ontologically; and the whole cosmos, as complete whole, hangs together, is "informed", by the powerful energy of one single higher monad that bestows its definitive intelligibility and power of action and reaction on everything below it. (*My Universe*, 1924, IX, 57 E; 85 F)

MONAD, GREAT *see also* **Monad; Noosphere**

First (poetic) description of the noosphere in which a lunar disc symbolizes unified human totality.

> As the particular consciousness of the monad spreads across the earth it seemed to me its disc was concentrated and illuminated at the same time as its path was more directly fixed on the zenith. The Great Monad has doubtless found one single, collective, human purpose of existence – and, each in its own degree, all individual effort has cooperated in this supremely vital task. (*The Great Monad*, 1918, XIII, 188 E; XII, 272–3 F)

MONOGENISM *see also* **Monophyletism**

Descent of humanity from a single couple. Non-teilhardian biological concept borrowed from theology. Antonym of polygenism.

> Once again, we must insist on the essential distinction between the notions (too often regarded as synonymous) of:
> • mono- and polygenism: one or more primitive couples;
> • mono- and polyphyletism: one or more branches (or phyla) at the base of humanity. (*Monogenism and Monophyletism*, 1950, X, 209 E; 247 F)

MONOPHYLETISM *see also* **Monogenism**

Descent of humanity from a single phylum. Biological concept adopted by Teilhard. Antonym of polyphyletism.

> Although the science of the human being can affirm nothing directly for or against monogenism (a single initial couple), it, on the other hand, seems to speak out decisively in favor of monophyletism (a single phylum). (*The Human Phenomenon*, 1938–40, I, 128 n. E; 208 n.1 F)

MONTINI

Giovanni Battista Montini, cardinal (1958) and future Pope Paul VI (1963–78).

> What you say about Mgr. Montini is (as they say in New York) "terribly exciting". What is the Holy Father waiting for to make him cardinal? (Letter to Jeanne-Marie Mortier, 25 January 1955, *Lettres à Jeanne Mortier*, 173)

MORALITY *see also* **Ethics; Metaphysics; Movement, Morality of; Spirit, Atomism of**

Ethics. Standard of human behavior based on what is considered ethically right or wrong.

(a) In a strict sense, "morality" is a coherent system of action that is:
- universal (governing all human activity) and
- categorical (entailing some form of obligation) . . .

If, in fact, "morality" (in a strict sense) implies coherence of action:
- either with a universal equilibrium (static morality)
- or with a universal movement (dynamic morality),

then it necessarily presupposes the categorical acceptance of a certain view of the world (either in equilibrium or in evolution). Otherwise, it remains "in the air", indeterminate. (*Can Moral Science dispense with a Metaphysical Foundation?*, 1945, XI, 130–1 E; 143–4 F)

(b) The world is ultimately constructed by moral forces; and, reciprocally, the function of morality is to construct the world. (*The Phenomenon of Spirituality*, 1937, VI, 105 E; 131 F)

(c) A system of sensitive and profound connections between spirit and matter exists deep within us. Not only, as christian moralists say, the one supports the other. But the one is born of the other . . . A new moral conception of matter appears underlying the religion (or morality) of spirit. (*The Evolution of Chastity*, 1934, XI, 68–9 E; 75 F)

(d) There is no longer a separate physical and moral domain around us. There is only a physico-moral domain. (*Sketch of a Personalistic Universe*, 1936, VI, 72 E; 90 F)

(e) Henceforth the highest morality is what will best develop the natural phenomenon to its upper limits. No longer to protect but to develop, by awakening and by convergence, the individual riches of the earth . . . Three principles define axiomatically the value of human actions:

(1) good is finally only that which contributes to the growth of spirit on earth;
(2) good (at least partially and basically) is that which causes a spiritual growth of the earth;
(3) best is finally that which assures the spiritual powers of the earth their highest development. (*The Phenomenon of Spirituality*, 1937, VI, 106 E; 132 F)

(f) Above all, the practical science of the human (morality, economics, politics, etc.) seems to have been conceived up to now as a problem of equilibrium: giving each individual his rights, his food, his land, etc. Clearly, it is now a question of recognizing a problem of energy . . . It seems to me moral is whatever uses and brings to blossom the grandeur that must be moralized. (Letter to Henri de Lubac SJ, 22 November 1936, *Lettres intimes*, 323)

MOVEMENT, MORALITY OF *see also* **Equilibrium, Morality of; Ethics**

Open morality. Opposite of closed morality. Dynamic ethics giving evolutionary and energetic meaning to virtues and values.

(a) Morality . . . must become dynamic if it is to seduce and save us. (*Christianity in the World*, 1933, IX, 103 E; 136 F)

(b) The moral world could well appear to the morality of balance ("closed morality") to be a definitely closed circle. This same world appears to the morality of movement ("open morality") to be a higher sphere of the universe, much richer in unknown powers and unsuspected combinations than the lower spheres of matter. (*The Phenomenon of Spirituality*, 1937, VI, 108 E; 134 F)

(c) A morality of balance can logically be agnostic and absorbed in possession of the present moment. A morality of movement is necessarily tilted towards the future and the pursuit of a God. (*The Phenomenon of Spirituality*, 1937, VI, 109 E; 135 F)

MOVEMENT, NEO-ANTHROPOCENTRISM OF

Humanity as the culmination or, rather, the leading shoot of evolution since humanity finds itself in evolution in line with the drift of complexity-consciousness (the axis of the progression of the cosmos and of humanity).

> To this neo-anthropocentrism of movement (man no longer the center but an arrow heading towards the center of the universe in process of concentration) there cannot fail to be objections. To establish my thesis I must, in fact, make three successive affirmations:
>
> - the critical nature of the reflective stage (part one);
> - the biological value of the social stage (part two);
> - the capacity of the universe to sustain the process of hominization and nourish it to the end without prematurely weakening or becoming exhausted (part three). (*The Singularities of the Human Species*, 1954, II, 209–10 E; 297 F)

MULTI- *see also* **Omni-**

Prefix. Multiform.

MULTIPLE, THE *see also* **Multiple, Pure; Multitude**

The primordial dust at the beginning of evolution that supports the process of centration and union.

> In its true and rightful sense, the multiple is understood as being convergent by nature. If it is to be reduced it must not be suppressed but extended beyond itself. (*The Road of the West*, 1932, XI, 46 E; 52–3 F)

MULTIPLE, PURE *see also* **Non-being, Creatable; Non-being, Physical**

Synonym of creatable or physical non-being. Appears in teilhardian views on creation.

> True non-being, physical non-being, non-being on the threshold of being, non-being where all possible worlds converge at their base, is pure multiple, multitude. (*The Struggle against the Multitude*, 1917, XII, 95 E; 132 F)

MULTITUDE *see also* **Multiple, The**

Synonym of the multiple with a more concrete nuance.

> Psychic simplicity, as we know it, is born of the multitude. It flourishes on organic complication, sovereignly numerous and sovereignly conquered. Proportional to the surmounted multitude that it shelters, it corresponds inexorably to the maximum tendency in being to become decomposed. (*The Struggle against the Multitude*, 1917, XII, 96–7 E; 133–4 F)

MUTATION *see also* **Orthogenesis**

Biological term representing the abrupt change of living forms that constitutes the birth of a new biological orientation through hereditary transmission. Teilhard integrates mutation in orthogenesis and gives it wider scope by extending it to every form of generation where the latter is seen as a small variation that cancels itself out as soon as it is born and does not involve any bifurcation of the phylum (species or sub-species).

> Life, too, must bud and divide; without this, the very existence of phyla would be inconceivable. That is to say, that unless the phenomena of continuous growth is considered directional (orthogenesis), we must pay increasing attention, we must recognize the place of movements of a wholly different nature: those of an abrupt change of form or, as we say, of mutation. Mutation or, as we have just explained, birth of new biological orientations that are beginning to be clearly observed by zoologists and botanists. (*The Movements of Life*, 1928, III, 144–5 E; 203 F)

MYSTICISM *see also* **Science**

1 In a universal sense, the search for the one behind the multitude, search that is born of a specific spiritual need and develops a specific art and science.

(a) By "mysticism" I understand here the need, the science and the art of attaining, at the same time and through each other, the universal and the spiritual. To become simultaneously and by the same act one with all, through liberation from all multiplicity or material gravity: here, deeper than any desire for pleasure, riches or power, is the essential dream of the human soul. (*My Fundamental Vision*, No. 32, 1948, XI, 199 E; 214 F)

(b) Mysticism is the supreme science and the supreme art, the only power capable of synthesizing the riches accumulated by other forms of human activity. (Letter to Henri Breuil, 9 September 1923, *Lettres inédites*, 143)

2 In a christian sense, search for union with God in Christ. Unity in its personal reality.

> The experiences described in this study are only an introduction to mysticism. Beyond the point at which I stopped, the being in whom the higher cosmic milieu is adequately personified reveals, as he wishes, the attraction of his face and heart. There are infinite degrees in this loving initiation of one

person in another unfathomable person. (*The Mystical Milieu*, 1917, XII, 147 E; 190–1 F)

3 In a teilhardian sense, final stage of the teilhardian vision (that, depending on the writing in question, consists of three or four stages). Constitutes the personalizing, christifying and evolutionary western mysticism as opposed to the eastern mysticism of dissolution and indeterminism.

(a) Re-write "My Universe". 1) Physics . . . 2) Dialectics: Point Omega, revelation, Christ Omega . . . 3) Metaphysics . . . 4) Mysticism . . . (Journal, 27 August 1947, in Claude Cuénot, *Nouveau Lexique*, 71)

(b) "Plan of work". In 4 parts: a physics . . . a dialectics . . . a metaphysics . . . a mysticism (evolutionary charity, Christ-the-humanizer, mysticism of the west). (Journal, 2 September 1947, in Claude Cuénot, *Nouveau Lexique*, 71)

(c) Science, in all probability, will be progressively more impregnated by mysticism (in order, not to be directed, but to be animated by it). (*My Universe*, 1924, IX, 83 E; 112 F)

MYSTICISM, EASTERN *see also* **East, Road of the**

Spirit as the opposite of matter. Unity obtained by a suppression of the multiple and dissolution of the person in an impersonal all, not by a personalizing convergence (unity of détente, not unity of tension).

(a) It would appear difficult to question the fact that, in many of its early manifestations, christianity appears as an offshoot of eastern mysticism. (*The Road of the West*, 1932, XI, 51 E; 57 F)

(b) Eastern mysticism. Mysticism only exists when the spirit seeks to resolve the opposition between unity and multiplicity, when there is hope of unity (no mysticism of pluralism). For eastern mysticism, the resolution of the multiple into the one is brought about by suppressing the multiple, unity having nothing in common with the multiple from which it must be separated (maya). The state of nirvana is an ecstasy of emptiness. (Lecture, 18 January 1933, Claude Cuénot, *Teilhard de Chardin*, 140 E; 174 F)

MYSTICISM, WESTERN *see also* **West, Road of the**

Spirit as the product of complexified matter. Unity obtained by a convergence of the multiple in a personal and personalizing focal point, not by suppression (unity of tension, not unity of détente).

(a) The time has certainly come when in the direction of a "pantheism of unification", a new mysticism at once fully human and fully christian can and must finally emerge at the antipodes of an outworn orientalism: the road of the west – the road of the world of tomorrow. (*Reflections on Two Converse Forms of Spirit*, 1950, VII, 227 E; 235–6 F)

(b) Western mysticism. It begins with the idea of the multiple being of a convergent nature, with elements capable of a gradual unity leading to unities of a higher order. Unity is obtained by the most thorough realization of things in themselves. (Lecture, 18 January 1933, Claude Cuénot, *Teilhard de Chardin*, 140 E; 175 F)

(c) The history of western mysticism could be described as a long attempt by christianity to recognize and separate, deep within itself, the eastern and western paths of spirituality: suppress or sublimate? Divinize by sublimation: it was on this side that the profound logic, the instinct of the nascent world, was to go. Divinize by suppression: it was in this direction that the accustomed ways of the ancient east were to push. (*The Road of the West*, 1932, XI, 51–2 E; 57 F)

N

NEGUENTROPY *see also* **Complexity, Principle of the greatest; Entropy**

Post-teilhardian expression (1967). Negative entropy. Anti-entropy. Opposite of entropy. Order, organization and improbability as opposed to disorder, disorganization and probability. Expresses the ascending primacy of life over entropy (where life is defined as a kind of anti- or negative entropy).

(a) The human, formerly seen as an anomaly in the universe, now tends to be seen as the extreme point so far attained in our experience by the combined process of corpuscular arrangement and psychic interiorization sometimes called "negative entropy" or "anti-entropy" or, more simply, evolution. (*The Contingence of the Universe*, 1953, X, 221 E; 265 F)

(b) We envisage as the basis of cosmic physics the existence of a sort of second entropy (or "anti-entropy") bearing, as the effect of chances taken, a part of matter in the direction of increasingly higher forms of structurization and centration. (*Transformation and Continuation in Man of the Mechanism of Evolution*, 1951, VII, 302–3 E; 317 F)

(c) By virtue of the statistical laws of probability . . . everything around us appears to be descending towards the death of matter; everything except life. (*The Movements of Life*, 1928, III, 149 E; 209 F)

NEO-

Prefix. Expresses the notion of transformation and renewal of ideas where the dynamic replaces the static, e.g. neo-act, neo-ahead, neo-anthropocentrism, neo-anthropogenesis, neo-anthropology, neo-barbarism, neo-brain, neo-buddhist, neo-catholic, neo-center, neo-cerebralization, neo-charity, Neo-Christ, neo-christianity, neo-christology, neo-circle, neo-condition, neo-consciousness, neo-convert, neo-copernican, neo-cortex, Neo-Creator, Neo-Cross, neo-darwinian, neo-darwinist, neo-detachment, neo-determinism, neo-dimension, neo-economy, neo-ego, neo-energetics, neo-energy, neo-envelope, neo-evil, neo-existential, neo-faith, neo-foundation, neo-galilean, neo-gene, neo-Gospel, neo-holiness, neo-human, neo-humanism, neo-humanity, neo-incarnation, neo-inquisition, neo-kosmos, neo-lamarckian, neo-leaf, neo-life, neo-light, Neo-Logos, neo-manicheanism, neo-manner, neo-marxist, Neo-Mass on the World, neo-materialist, neo-matter, neo-metaphysics, neo-milieu, neo-moral, neo-moslem, neo-movement, neo-multiple, neo-mysticism, neo-notion, neo-ontology, neo-paganism, neo-pantheism, neo-parameter, neo-paternity, neo-pelagianism, neo-personality, neo-plan, neo-pleromization, neo-property, neo-protestant, neo-psycho-analysis, neo-psychology, neo-redemption, neo-reform, neo-religion, Neo-Sacred Heart, neo-sanctity, neo-sense, neo-sin, neo-socialization, neo-space, neo-sphere, neo-spirit, neo-spirituality, neo-state, neo-step, neo-supernatural, neo-taste, neo-theology, neo-time, neo-type, neo-value, neo-work, neo-zest . . .

NEO-CHRIST

Christ to the extent that his cosmic dimension and completion of the Mystical Body appear profoundly renewed in human consciousness within the framework of cosmogenesis.

> No longer "The Exercises" but (very humbly and because I am compelled from within) "My Exercises". The result of others, naturally. But with "The" in addition. The neo-foundation, I repeat, Neo-Christ, Neo-Cross – the new Heart. And this, I repeat, in complete submission to the Church; since the Neo-Christ can only be a Trans-Christ. (Retreat, Purchase, 23 June 1952, in Claude Cuénot, *Nouveau Lexique*, 134)

NEO-CHRISTIANITY

Not a new christianity but a renewed christianity that has assimilated the senses of evolution, the human and the ahead.

> What do we find if our minds can embrace simultaneously both contemporary neo-christianity and contemporary neo-humanism, and so suspect and then accept as proven that the Christ of Revelation is none other than the omega of evolution? (*The Christic*, 1955, XIII, 92 E; 106 F)

NEO-HUMANISM *see also* **Cosmogenesis, Humanism of**

> It is in the direction of a dynamic and progressive neo-humanism (one, that is, that is based on humans having become conscious of being the responsible axis of cosmic evolution) that a mysticism of tomorrow is beginning to assert itself as the answer to the new and constantly increasing needs of anthropogenesis. (*My Fundamental Vision*, No. 36, 1948, XI, 202–3 E; 217 F)

NEO-LOGOS *see also* **Christ, Cosmic; Christ, Universal**

By analogy with the Alexandrian logos that represented the spirit that gave form statistically to the cosmos and that was integrated in the Christ of the Gospels, the neo-logos represents the principle of a spiritualizing genesis and an evolutionary convergence whose initial focal point and ultimate center is found in the Cosmic Christ, who is none other than the

Christ of the Gospels whose cosmic and evolutionary dimensions have been discovered by human consciousness.

> In the first century of the Church, christianity made its definitive entry into human thought by boldly identifying the Jesus of the Gospels with the Alexandrian Logos. Why can't we see the logical continuation of the same line of thinking and the prelude to the success found in the instinct that is now impelling the faithful, after two thousand years, to adopt the same approach, no longer with the ordinating principle of the stable Greek kosmos, but with the Neo-Logos of modern philosophy – the evolutionary principle of a universe in movement? (*Christianity and Evolution*, 1945, X, 180–1 E; 211 F)

NEO-MATTER *see also* **Matter, Secondary**

(a) Here, in its turn, is the long series of vital declines: the exhaustion and aging of the races, their collapse into lassitude, their encrustation beneath social envelopes that have become gilded and sterile castes, their stiffening under collective and individual routine; and here, finally, above this neo-matter in constant process of forming and rejecting, vast and ancient matter reappears. As imponderable in appearance as the inorganic world beneath the impassive mask of statistical laws, the determinism of large numbers and the painful friction of unorganized masses cover and level the quivering inner sheet of the noosphere. (*Hominization*, 1923, III, 70–1 E; 101 F)

(b) In the degree to which every aggregation of consciousness is not harmonized it is automatically enveloped at its own level by a veil of "neo-matter" – matter, which is the tangential face of every living mass in process of unification. (*The Human Phenomenon*, 1938–40, I, 182 E; 284–5 F)

NEO-MODERNISM

Not a new form of modernism but the expression of a legitimate wish to understand traditional christianity in the context of cosmogenesis and the modern world. Rarely-used expression.

> The only case of heresy involving neo-modernism is that of the "Ultra-Christ". Christ cannot be greater and more beautiful than himself. So . . . (in the case of Christ the ultra is true). (Journal, Epiphany, 6 January 1945, in Claude Cuénot, *Nouveau Lexique*, 136)

NEO-PANTHEISM, HUMANIST *see also* **Pantheism**

> I shall say that in my view a preliminary examination is sufficient to reduce possible types of belief to three. The group of eastern religions, the humanist neo-pantheisms and christianity: these are the directions between that I might hesitate were I to find myself (as I imagine hypothetically here) in the position of really having to chose my religion again. (*How I Believe*, 1934, X, 121 E; 141–2 F)

NEO-SPIRIT, NEO-SPIRITUALITY

In a universe whose convergent nature has been recognized, a neo-spirituality for a neo-spirit. (*The Atomism of Spirit*, 1941, VII, 57 E; 63 F)

NEWMAN

John Henry Newman (1801–90), cardinal and theologian. Major influence on Vatican II (1962–5). Teilhard greatly admired Newman (see also Henri de Lubac SJ, *Lettres intimes*, 407).

(a) When de Lubac feels he has rested enough (that I imagine is why he is at Gap!!!) you and I should press him to write his "Apologia" (in the Newman sense). (Letter to Bruno de Solages OCarm, 2 February 1952, *Lettres intimes*, 406)

(b) Why (as I've often said to him) doesn't de Lubac do a Newman and write his "Apologia" (not a polemic, of course, but the full manifestation of his full thought)? (Letter to Bruno de Solages OCarm, 26 September 1952, *Lettres intimes*, 414)

(c) I have been reading Thureau-Dangin's *Newman catholique* . . . I feel more than ever in sympathy with the great Cardinal, so undaunted, so firm of faith, so full, as he says of himself, "of life and thought" – and, at the same time, so thwarted. (Letter to Marguerite Teillard-Chambon, 22 July 1916, *Genèse d'une pensée*, 145; *The Making of a Mind*, 114)

NEW RELIGION

Some of his friends and admirers were inclined to see Teilhard ill at ease in the Church of his day. One remarked, "Your religion is admirable but it is not the catholic religion." Teilhard replied, "Do you really think me stupid enough to want to start a new religion? Or believe myself a second Jesus Christ?" (see also René d'Ouince SJ, "L'Obéissance dans la vie du Père Teilhard de Chardin," in *L'Homme devant Dieu, Mélanges offerts au Père Henri de Lubac*, Aubier, 1964, 343).

NICÆA, NEW *see also* Christ, Cosmic; Church

Council anticipated by Teilhard. Only partly realized in the Second Vatican Council (1962–5) that dealt with the relations, not between Christ and the universe, but between the Church and the modern world (see also Henri de Lubac SJ, *Teilhard Posthume*, 145–6; *Lettres intimes*, 430 n.2).

(a) It seems we are now reliving after 1,500 years the great conflicts with arianism – with the big difference that we are now concerned with defining the relations, not between Christ and the Trinity, but between Christ and a universe that has suddenly become fantastically large, formidably organic and more than probably poly-human (n thinking planets – millions perhaps). And if I may express myself brutally (but expressively) I see no valid or constructive way out of the situation except by making through the theologians of a new Nicæa a sub-distinction in the human nature of Christ between a terrestrial nature and a cosmic nature. (Letter to André Ravier SJ, 14 January 1955, *Lettres intimes*, 452)

(b) The eventuality of the plurality of thinking planets is no longer what you call "a possibility" . . . It has become a positive probability (and even a very real probability)(something recently noted in passing by a leading American astronomer) . . . I am more than ever convinced that we shall need, sooner or later, a new Nicæa that will define the cosmic face of the incarnation. (Letter to Bruno de Solages OCarm, 16 February 1955, *Lettres intimes*, 459)

NICHOLAS OF CUSA *see also* **Angela di Foligno**

Nikolaus von Kues or Nicolaus Cusanus (1401–64), cardinal, statesman and mystic. Rejected geocentrism and taught the plurality of inhabited worlds (see also Henri de Lubac SJ, *Lettres intimes*, 40). Major influence on Teilhard's spiritual and intellectual development. Teilhard would agree with Nicholas: "the divine is the enfolding of the universe and the universe is the unfolding of the divine" (*De Conjecturis*, 1442–3).

We need only to observe the rising rise of consciousness around us to see its presence everywhere. Plato had already sensed this and given it immortal expression in his Dialogues. Later on, with thinkers like Nicholas of Cusa, medieval philosophy technically returned to the same idea . . . (*The Human Phenomenon*, 1938–40, I, 188 E; 294 F)

NON-BEING *see also* **Being**

Limited multiplicity. Total exteriority.

(a) Complete exteriority or total "transience", like absolute multiplicity, are synonyms of non-being. (*My Universe*, 1924, IX, 46–7 E; 75 F)

(b) To create, even when we use the word "omnipotence", should no longer be understood as an instantaneous act but as a process or controlled movement of synthesis. Pure act and "non-being" are as totally opposed as perfect unity and pure multiple. (*Christology and Evolution*, 1933, X, 82–3 E; 101 F)

NON-BEING, CREATABLE *see also* **Non-being, Positive**

By the very fact of his being centered on himself in order to exist, the first being ipso facto stimulates another form of opposition, no longer in his heart but at the very opposite pole to himself (phase three). The self-subsistent unity, at the pole of being, and, as a necessary consequence, the multiple all around it, on the perimeter: the pure multiple . . . or "creatable non-being" that is nothing – but that is, nevertheless, a possibility of being, a plea for being, by a passive potentiality of arrangement (that is, of union). (*My Fundamental Vision*, No. 28, 1948, XI, 194 E; 209 F)

NON-BEING, PHYSICAL *see also* **Multiple, Pure; Non-being, Positive**
Opposite of conceptual non-being that is both absurd and unthinkable. Represents the limited multitude.

> Pure non-being is an empty concept, a pseudo-idea. True non-being, physical non-being, non-being on the threshold of being, that on which all possible worlds converge at their base, is pure multiple, is multitude. (*The Struggle against the Multitude*, 1917, XII, 95 E; 132 F)

NON-BEING, POSITIVE *see also* **Non-being, Creatable**
Second phase of theogenesis. Possibility of creation. Represents the primordial multiple that arrives at the point of being at the same time as it begins to become centered through the effect of the creative union of God. Synonym of creatable non-being.

(a) God envelops himself in participated being by evolutive unification of the pure multiple ("positive non-being") born – in a state of absolute potentiality – through antithesis to pre-posited Trinitarian unity: creation. (*Christianity and Evolution*, 1945, X, 178 E; 209 F)
(b) Positive non-being . . . Nothing can be positive that has not already been subjected to the first effects of union. (*Creative Union*, 1917, XII, 162–3 E; 208–10 F)

NOODYNAMICS
Energetic conditions for the development of reflective thought in relation to the cosmos.

> Noodynamics: dynamism of spiritual energy, dynamics of the spirit. I have ventured to use this neologism because it is clear, expressive and convenient; also because it affirms the necessity for incorporating human psychism, thought, in a true "physics" of the world. (*The Human Rebound of Evolution*, 1947, V, 206 E; 267 F)

NOOGENESIS *see also* **Noosphere; Psychogenesis**
Third phase of evolution. Evolution of the mind and spirit (reflective life). Noogenesis represents the movement of the universe to the extent that, by a gradual process of concentration of increasingly arranged and better centered systems, it leads to the emergence of the noosphere at the term of a drift of complexity-consciousness. Teilhardian "keyword".

> In us and through us noogenesis constantly continues to rise. We have recognized the main characteristics of that movement: the drawing together of grains of thought; the synthesis of individuals and synthesis of nations or races; the necessity of an autonomous and supreme personal focal point to join the elementary personalities together in an atmosphere of active sympathy, without deforming them. All this, once again, from the combined effect of two curves: the sphericity of the earth and the cosmic convergence

of mind – in conformity with the law of complexity and consciousness. (*The Human Phenomenon*, 1938–40, I, 205–6 E; 319–20 F)

NOOSPHERE

Often called the "sphere of mind". More correctly, the "sphere of mind-and-spirit". The French word "esprit" follows the Greek "noos" in covering both "spirit" and "mind" in English. The spiritual (or thinking) layer of the world, a new biological kingdom, an organic and specific whole in process of unanimization, distinct from the biosphere, the non-spiritual (or non-thinking) layer of the world. Expression first used in *Hominization* (1925). Represents as important an evolutionary leap forward as the atmosphere and hydrosphere. Teilhardian "keyword".

(a) Although this view may seem at first sight both exaggerated and fantastic, what we now propose is to consider the thinking envelope of the biosphere as being of the same order of zoological magnitude (or, if you prefer, of telluric magnitude) as the biosphere itself. The more we think about it, the more this extreme solution seems the only honest one. Unless we give up all attempts to restore humans to their place in the general history of the earth as a whole, without damaging them or disorganizing it, we must place them above it, without, however, uprooting them from it. And this amounts to imagining, in one way or another, above the animal biosphere a human sphere, the sphere of reflection, of conscious invention, of the conscious unity of souls (the noosphere, if you like) and conceiving, at the origin of this new entity, a phenomenon of special transformation affecting pre-existent life: hominization. (*Hominization*, 1923, III, 63 E; 91–2 F)

(b) Around us, tangibly and materially, the thinking envelope of the earth – the noosphere – is multiplying its internal fibers and tightening its network; and, simultaneously, its internal temperature is rising and its psychism is mounting. (*The Planetization of Humanity*, 1945, V, 132 E; 167 F)

NUMBERS *see also* Numbers, Large; Numbers, Law of large

The abyss of numbers – a terrifying floodtide, all around us, of bodies and corpuscles. (*A Phenomenon of Counter-Evolution in Human Biology*, 1949, VII, 185 E; 192 F)

NUMBERS, LARGE *see also* Chance

Category of teilhardian thought that provides a statistical interpretation of determinism.

I cannot admit that the universe is a failure. This privilege (the assurance of success) may be due either to a providential transcendent action or to the influence of a spiritual energy immanent in the world (some soul of the world), to a sort of infallibility that, though not accorded to isolated attempts, attaches to indefinitely multiplied attempts ("the infallibility of large numbers") or, again, it may, more probably, be derived from the hierarchically ordered action of these three factors at the same time. (*My Universe*, 1924, IX, 40–1 E; 69 F)

NUMBERS, LAW OF LARGE

Law that states that a sufficiently large number of cases will permit the appearance of exceptional opportunities that will favor the development of a higher stage of evolution. Statistical law transposed into an evolutionary dimension.

(a) Physico-chemical matter is an abstraction (in the strictest sense of the word) obtained by isolating the cosmic elements from everything that is a higher unification at a certain level . . . This level is determined by the presence in the multiple of a sufficient proportion of determinisms that is due either to a statistical effect of "large numbers" or to an automatism appearing in the monads in a secondary mode. (*The Names of Matter*, 1919, XIII, 229, 238 n.3 E; XII, 453 n.4 F)

(b) The human mass, as a whole, still obeys the same laws of large numbers that allow science to treat gaseous masses or any other particular grouping as a mechanical thing. To a sufficiently distant observer the sum total of our free choices would appear overlaid by determinisms. (*The Names of Matter*, 1919, XIII, 230 E; XII, 455 F)

(c) Physical determinisms ("laws") are simply the effects of large numbers, that is, of materialized freedom. This statistical materialization of the "Weltstoff" is, of course, most marked in the zone of "fragmentary centers" (infinitely numerous and infinitesimally spontaneous); but it is still perceptible among centers of a higher order and even in the noosphere . . . There is no proof that, in accordance with some law of large numbers, there may not be many other obscure stars and many earths like ours already scattered or still expected among the galaxies. (*Centrology*, 1944, VII, 125, 127 E; 132, 134 F)

O

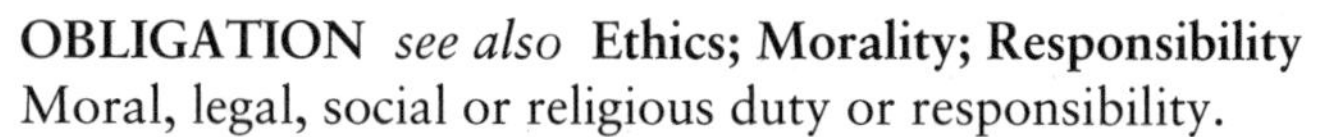

OBLIGATION *see also* **Ethics; Morality; Responsibility**

Moral, legal, social or religious duty or responsibility.

(a) One of the most important aspects of hominization . . . is the accession of biological realities (or values) to the domain of moral realities (or values). With the human and in the human evolution has become reflectively conscious of itself . . . In a spiritually evolutionary perspective . . . the initial basis of obligation is the fact of being born and developing in function of a cosmic stream. We must act in a certain way because our individual destinies depend on a universal destiny. In origin, duty is nothing other than the reflection of the universe in the atom. (*The Spirit of the Earth*, 1931, VI, 29 E; 36 F)

(b) Henceforth we must take account, more explicitly, of the obligations of humans towards the collective and even towards the universe: political duties, social duties, international duties – cosmic duties (one might say) of the first order that include the law of work and the law of research. A new horizon of responsibilities is being opened up to our contemporaries . . . (*Note on the Presentation of the Gospel in a New Age*, 1919, XIII, 220 E; XII, 411 F)

OMEGA *see also* **Christ-Omega**

Teilhardian "keyword".

1 From the point of view of emergence, center defined by the final concentration of the noosphere on itself. Natural point of convergence of humanity and the whole cosmos. Term of social and spiritual maturation of the earth.

2 From the point of view of transcendence and pre-existence, one of the two apparent poles of God, namely, God the end of creation acting through the mediation of Christ-Omega. In reality, the two poles of God, alpha and omega, beginning and end, coincide in the divine unity and eternity (see also the dynamic Omega of Teilhard and the static Omega of the Spiritual Exercises, Jacques Laberge SJ, *Pierre Teilhard de Chardin et Ignace de Loyola*, 193).

(a) For the noosphere to be actual and real, the center must be actual and real. To be extremely attractive, omega must already be supremely present. (*The Human Phenomenon*, 1938–40, I, 192 E; 299–300 F)

(b) Omega . . . must be presented as personal, individual, already partially actual and partially transcendent. (*Centrology*, 1944, VII, 112 E; 118 F)

(c) The noosphere physically requires, for its maintenance and functioning, the existence in the universe of a true pole of psychic convergence: a center different from all other centers that it "super-centers" by assimilation; a personality distinct from all other personalities that it perfects by uniting with them. The world would not function if there did not exist, somewhere ahead in time and space, "a cosmic Point Omega" of total synthesis. (*Human Energy*, 1937, VI, 145 E; 180 F)

(d) a last and supreme definition of the Omega Point: focus both unique and complex in which, bound together by the christic person, three interlocking centres are ever more deeply revealed: externally, the immanent ("natural") summit of the humano-cosmic cone; further in, towards the middle, the immanent ("supernatural") summit of the "ecclesiastic" or christic cone; and finally, at the very heart, the transcendent centre, triune and divine. The complete Pleroma coming together under the mediating action of Christ-Omega. (*Outline of a Dialectic of Spirit*, 1946, VII, 148 E; 156 F)

(e) In the perspectives of a noogenesis, time and space become truly humanized – or rather, super-humanized. Far from being mutually exclusive, the universal and the personal (that is, the "centered") grow in the same direction and culminate in one anther at the same time. It is therefore an error to look for the prolongation of our being and of the noosphere in the direction of the impersonal. The future of the universe can only be hyper-personal – in the Omega Point. (*The Human Phenomenon*, 1938–40, I, 184–5 E; 288–9 F)

(f) Comparable in every way to the Point Omega that our theory led us to foresee, Christ (provided he reveals himself in the full realism of his incarnation) tends to produce exactly the spiritual totalization we expected. (*Human Energy*, 1937, VI, 155 E; 192 F)

(g) For millions and millions of believers (among the most aware human beings) Christ, since he appeared, has never ceased re-emerging after each crisis of history with greater presence, greater urgency and greater penetration than ever before. What does he still lack if he is to show himself once again to our modern world as the "new God" we expect? Two things in my view; and two things only. First, in a universe where we can no longer seriously consider thought an exclusively terrestrial phenomenon, he can no longer be restricted *constitutionally* in his operation to a simple "redemption" of our planet. And second, in a universe where we can now see everything co-reflected along a single axis, he can no longer be offered for our worship (as the result of a subtle and pernicious confusion between "super-natural" and "extra-natural") as a distinct and rival summit to the summit to which the biologically extended path of anthropogenesis is leading us. (*The God of Evolution*, 1953, X, 241–2 E; 289–90 F)

OMEGA, POINT *see* **Omega**

OMEGAGENESIS *see also* **Christogenesis**
Fourth (and final) phase of evolution. Process of convergence on Point Omega. Synonym of christogenesis.

To communicate with christogenesis (omegagenesis). (Retreat notes, 1941, in Jacques Laberge SJ, *Pierre Teilhard de Chardin et Ignace de Loyola*, 159)

OMEGALIZATION *see also* **Omegalize**

Progressive transfiguration of the universe that is increasingly centered on its christic nucleus through the conjugated action of grace and human effort.

(a) In fact, all who succeed in seeing existing, not only in the immense and the infinitesimal, but partly in the complex, a way of acting capable of synthesizing and transforming all other forms of action: the specific action of sustaining and promoting, in and around themselves – on the whole surface and in the depths of the real, the unification of the universe (and consequently its awareness) on its profound center; the total and totalizing action (if I may use the word – I can find no other) of "omegalization". (*The Atomism of Spirit*, 1941, VII, 55–6 E; 62–3 F)

(b) Omegalization = to communicate with becoming (emergence). (Retreat notes, 1940, in Jacques Laberge SJ, *Pierre Teilhard de Chardin et Ignace de Loyola*, 159)

OMEGALIZE *see also* **Omegalization**

All around us the universe, reduced to its eu-centric portion (to its essence), is cautiously re-shaping itself, "omegalized" grain by grain, through death. (*Centrology*, 1944, VII, 122 E; 129 F)

OMNI- *see also* **Pan-**

Prefix. Homogeneous. Expresses the notion of orientation towards totality and totalization, e.g. omni-action, omni-potence, omni-presence, omni-science, omni-sufficiency, omni-sufficient . . .

ONE, THE *see also* **Union, Richness of**

Concept representing an act of synthesis of the multiple at each stage of evolution. Opposed to the one of static metaphysics.

When . . . we admit the idea of a genesis or fulfillment of the one from the elements of the world, these elements, while remaining, because of their temporary condition of disorder . . . the source of all sin and evil, no longer have need of a prior evil to explain their appearance and initial distribution. (*The Road of the West*, 1932, XI, 57 E; 62 F)

OPIUM

Metaphor.

(a) Religion can become an opium. It is too often understood as a simple antidote to our suffering. Its true purpose is to sustain and to spur on the progress of life. (*The Spirit of the Earth*, 1931, VI, 44 E; 53 F)

(b) The Cross is not a shadow of death but a symbol of progress. Christianity does not dispense the opium of a defeatist passivity but the lucid intoxication of a magnificent reality to be discovered through an advance across the broad front of the universe. (*Christianity in the World*, 1933, IX, 108 E; 140–1 F)

(c) There is a communion (the true communion) with God through the world . . . Such a christianity is still in reality the true evangelism since it represents the same force applied to the elevation of humanity above the tangible through common love. Yet, at the same time, this evangelism has none of the taint of opium that we are so bitterly accused (and with some reason) of dispensing to the masses. (*Christology and Evolution*, 1933, X, 93 E; 111 F)

(d) It is disquieting to see here and in Rome the growing number of (emotional "Marian") rallies crying "death to materialism" without understanding that for those of us (who believe in cosmogenesis) the . . . only way of defeating communism is to present Christ as he should be: not as an opium (or its derivative) but as the essential mover of a hominization that can only be completed energetically in a world that is "amorized" and open at the summit. (Letter to Pierre Leroy SJ, 14 October 1952, *Lettres familières*, 164; *Letters from My Friend*, 148)

ORGANICITY

Naturally harmonious character of a material or spiritual whole, from the space–time of cosmogenesis to the Mystical Body.

(a) At this degree of generality, where evolution simply means the organicity of the stuff of the universe (temporal organicity combined with spatial organicity), at this degree, I repeat, it is not enough to speak of certitude. We must speak of "evidence". (*The Degrees of Scientific Certainty*, 1946, IX, 193 E; 246 F)

(b) Undoubtedly . . . it was the experience of war that made me aware of a sort of sixth sense of this still comparatively rare gift or faculty of perceiving, without seeing, the reality and organicity of collective magnitudes. (*The Heart of Matter*, 1950, XIII, 31 E; 41 F)

ORGANO-PSYCHIC

Solidarity comparable to that of an organism and, by extension, a certain stage of psychic emergence. Characterizes a level of phenomena where psychism emerges in conjunction with organic complexity.

The rise of civilization is nothing but the organo-psychic aspect adopted by a colossal biological operation never before attempted in nature: the arrangement around itself . . . of a vast group of living groups, I mean, of a whole phylum (and a phylum of planetary extension). (*The Phyletic Structure of the Human Group*, 1951, II, 156 E; 216 F)

ORTHODOXY *see also* **Exercises, Spiritual**

Ignatius did not see what could (and must, in fact,) "equip" his Society was not static unity (conformism) in doctrine but doctrinal "pioneerism" following "the christic axis" and "the pleromic axis" (static orthodoxy versus dynamic orthodoxy . . .). (Journal, 13 March 1955, in Jacques Laberge SJ, *Pierre Teilhard de Chardin et Ignace de Loyola*, 200–1)

ORTHO-ELECTION *see also* **Ortho-selection**

Active form of orthogenesis dominated by conscious or unconscious inventiveness. Opposed to ortho-selection.

By nature, this phenomenon of initial reflection has two main effects:

(1) to ensure in the human the gradual transition of life from a state of sustained evolution (ortho-selection) to a state of self- or directed evolution (ortho-election);

(2) to secure the predominance in the thinking layer of the earth of the convergent forces of hominization over the divergent forces of speciation (or phyletization). (*A Major Problem for Anthropology*, 1951, VII, 316 E; 330 F)

ORTHOGENESIS *see also* **Canalization, Mutation by; Cephalization; Ortho-election; Ortho-selection**

Direction and orientation in evolution.

1 In a biological sense, cumulative series of small anatomical (and psychic) mutations oriented in the same direction, thereby constituting a phenomenon of continuous growth in the same direction, something that differentiates it from irreversibility that can contain phenomena of regression or detours within a major movement of the whole.

2 In a phenomenological sense, fundamental drift according to which the stuff of the universe seems to shift towards corpuscular states of increasingly complex material arrangement – states that, psychically, are increasingly interiorized; drift that appears in higher forms of life through cephalization and in human beings through the appearance of reflection.

(a) Orthogenesis . . . is a perfectly simple and obvious notion – since it does no more than express the indisputably "fibrous" and "radiated" aspects that everyone agrees we see in the biosphere. The real difficulty – and the real interest – begins when, taking one more step forward, we find ourselves asking:

(1) whether (and to what extent) the incontestably directed additive quality of "speciating" mutations in certain privileged directions (phyla) is based:

- either on a particular structure of the external milieu within which successive mutations take place: passive orthogenesis or ortho-selection;
- or, on the contrary, on an internal "preference" (conscious or unconscious) of the living being to follow one direction as opposed to another: active orthogenesis or ortho-election.

(2) whether, under the generic term "orthogenesis" or phyletization, two equally important and equally profound processes have not been confused by chance:

- one process of specialization, leading to the birth of increasingly divergent and differentiated forms;
- the other process of complexification (or complexity), producing, along all the azimuths of specialization (with comparative success in each case), increasingly centered and cerebralized zoological types. (*Note on the Reality and Significance of Human Orthogenesis*, 1951, III, 250–1 E; 355–6 F)

(b) Whether it wants to or not, paleontology is, and can only become more, the science of orthogenesis. (*A Defense of Orthogenesis*, 1955, III, 273 E; 390 F)

(c) "Orthogenesis" . . . is an essential and irreplaceable word to indicate and affirm the obvious property living matter possesses of forming a system "within which terms succeed each other experimentally according to constantly increasing values of centro-complexity". (*The Human Phenomenon*, 1938–40, I, 65 n. E; 114–15 n.1 F)

According to Teilhard an examination of the evolutionary process as a whole shows the main features that occur at both the micro and the macro levels. These features are:

(a) an increasing complexification of forms from the earliest to the latest (or from the "lowest" to the "highest"), and

(b) an increase in consciousness also from the earliest to the latest. (P. G. Fothergill, Teilhard and the Question of Orthogenesis, in *Evolution, Marxism and Christianity*, 40)

The evidence of progress and directionality in biological evolution is clear enough if the living world is considered as a whole . . . Evolution has produced organisms with highly developed nervous systems, that convey to them information about the states of their environments. To some extent, such organisms can dominate their environments, instead of being dominated by the latter . . . It is the totality of evolution that occupies Teilhard's attention almost exclusively. The only particular evolutionary line that interests him is that of man, and this because he believes that in man evolution as a whole is, at is were, brought into focus . . . Orthogenesis is a hypothesis that endeavors to explain what causes evolution, rather than a summary description of the evolutionary history of the world. (Theodosius Dobzhansky, *The Biology of Ultimate Concern*, 119)

ORTHOGENESIS, ACTIVE *see also* **Ortho-election**

The real difficulty – and the real interest – begins when, taking one more step forward, we find ourselves asking . . . whether (and to what extent) the incontestably directed additive quality of "speciating" mutations in certain privileged directions (phyla) is based:

- either on a particular structure of the external milieu within which successive mutations take place: passive orthogenesis or ortho-selection;
- or, on the contrary, on an internal "preference" (conscious or unconscious) of the living being to follow one direction as opposed to another: active orthogenesis or ortho-election. (*Note on the Reality and Significance of Human Orthogenesis*, 1951, III, 250 E; 356 F)

ORTHOGENESIS, BASIC *see also* **Cephalization, Law of; Greater Consciousness, Evolution of**

Drift of complexity-consciousness showing itself, on the level of life, in increasingly complex systems orientated towards maximum cerebration.

We insisted earlier on the general orthogenesis of corpusculization, on "basic orthogenesis", that, we said, draws all matter towards the more complicated and the more conscious. (*The Singularities of the Human Species*, 1954, II, 219 E; 308 F)

ORTHOGENESIS, PASSIVE *see also* **Orthogenesis, Active; Ortho-selection**

ORTHO-SELECTION *see also* *Ortho-election; Orthogenesis, Active*

Passive form of orthogenesis dominated by the selective influence of the external environment. Opposed to ortho-election.

By nature, this phenomenon of initial reflection has two main effects:

(1) to ensure in the human the gradual transition of life from a state of sustained evolution (ortho-selection) to a state of self- or directed evolution (ortho-election);

(2) to secure the predominance in the thinking layer of the earth of the convergent forces of hominization over the divergent forces of speciation (or phyletization). (*A Major Problem for Anthropology*, 1951, VII, 316 E; 330 F)

OVERTHROW *see* **Reversal**

P

PAIN *see also* **Cross, Sense of the; Suffering**

Every aspect of human suffering.

> Pain of personalization . . . Pain of plurality . . . Pain of differentiation . . . Pain of metamorphosis. (*Sketch of a Personalistic Universe*, 1936, VI, 84–7 E; 105, 107, 108 F)

PAN- *see also* **Omni-**

Prefix. Expresses the notion of orientation towards totality and totalization, e.g. pan-amorization, pan-analyzable, pan-attraction, Pan-Christ, pan-christic, pan-christifying, pan-christism, pan-cohesion, pan-communion, pan-consecration, pan-consummator, pan-contact, pan-continental, pan-corpuscular, pan-determinism, pan-element, pan-energetics, pan-groping, pan-human, pan-humanizing, pan-integration, pan-interliaison, pan-liaison, pan-organized, pan-polarization, pan-polyphyletic, pan-presence, pan-psychism, pan-reflective, pan-relativity, pan-sensorial, pan-structure, pan-telepathy, pantheism, pantheisticity, pan-union, pan-urgency, pan-inside . . .

PAN-AMORIZATION *see also* **Amorization**

Universal amorization spreading from a final center of personalization.

> Christianity, far from losing its primacy in the great religious ferment unleashed by the totalization of the modern world, regains and consolidates its axial and directional position as the spearhead of human psychic energies: provided we give sufficient attention to its extraordinarily significant power of "pan-amorization". (*The Christic*, 1955, XIII, 88 E; 102–3 F)

PAN-CHRISTIC *see also* **Pan-christism**

> Although I had not taken sufficient account, at first, of the bridge between my love of Jesus and my love of things provided by this thoroughly christian attitude, I have never, from the earliest years of my religious life, ceased abandoning myself willingly to this active feeling of communion with God through the universe. And it was the decisive emergence of this "pan-christic" mysticism, definitely matured in the heady atmosphere of Asia and the Great War, that was to be reflected, in 1924 and 1927, in "The Mass on the World" and "The Divine Milieu". (*The Heart of Matter*, 1950, XIII, 47 E; 59 F)

PAN-CHRISTISM *see also* **Personal-universal**

Christ, universal and personal, alpha and omega, all are given equal value in the teilhardian vision of cosmogenesis that sees the universe fully completed in a final center through the spiritualizing and unifying effects of evolution and the intimate union of alpha and omega brought about through the revelation of its temporal stuff.

> Necessarily, in a system of creative union, it is not only the universe, it is God himself who is "christified" in omega at the upper limits of consciousness. In other words, an "evolved" monotheism around that the very best of the spiritual energies of the earth certainly appear to be concentrating, is going to be completed logically and biologically in the direction of some sort of pan-christism. (*The Heart of Matter*, 1950, XIII, 55 E; 67 F)

PAN-COMMUNION

Expression representing, not the real presence in the Eucharist, but the participation of the incarnate Christ through the all.

> I have never, from the earliest years of my religious life, ceased abandoning myself willingly to this active feeling of communion with God through the universe. And it was the decisive emergence of this "pan-christic" mysticism, definitely matured in the heady atmosphere of Asia and the Great War, that was to be reflected, in 1924 and 1927, in "The Mass on the World" and "The Divine Milieu . . . " Undoubtedly, as a basis for the pan-communion that obsessed and intoxicated me at that time, I already had at my disposal a world whose elements were organically woven and whose layers were organically linked. But this ambient organicity, the specific basis of the christic diaphany, still only existed, in my mind and in my eyes, in what we might call a diffused state. (*The Heart of Matter*, 1950, XIII, 47–8 E; 59 F)

PAN-ENERGETICS *see also* **Energetics**

Generalized science of energy.

PANENTHEISM *see also* **Pantheism, Christian**

Lit. "God-in-all". God is in all and all is in God. Antonym of pantheism – God is all and all is God. It describes the intimate interdependence between God and the universe while avoiding the identification of the universe with God (pantheism). Its roots are to be found in Plato. Probably first used by Karl Christian Friedrich Krause (1781–1832), it has since been given expression by various twentieth-century thinkers, e.g. Alfred North Whitehead, Charles Hartshorne, Pierre Teilhard de Chardin . . . Teilhard does not use the term *expressis verbis*. He calls it christian pantheism.

PAN-REFLECTIVE

Final stage of the centric when total radial (or spiritual) energy, fully centered, transfigures by adoption every manifestation of tangential (or physical) energy.

> The secret and mainspring of my spiritual drive will have been to see that, underlying this external envelope of the phenomenon (but in genetic continuity with it), there stretched another domain (no longer tangential but centric) where a second type of energy (no longer electro-thermodynamic but spiritual) radiating from a first divisible point in ascending order, in three increasingly interiorized zones:
>
> - first, the zone of the human (or the reflective);
> - second, the zone of the ultra-human (or the co-reflective);
> - third, the zone of the christic (or the pan-reflective). (*The Stuff of the Universe*, 1953, VII, 375–6 E; 398 F)

PANTHEISM *see also* Panentheism

In a teilhardian sense, the universe itself is not an immanent and impersonal God but capable, through self-transfiguration, of progressively becoming the divine milieu where the presence of a transcendent and personal God is made increasingly real so that union with the all becomes the way to mystical union with God.

> Finally, to end once and for all the fears of "pantheism" constantly promoted about evolution by certain upholders of the traditional spiritual view, in the case of a convergent universe as I have presented it, who cannot see that, far from being born of the fusion and confusion of the elementary centers it assembles, the universal center of unification (precisely in order to fulfill its motive, collecting and stabilizing function) must be conceived as pre-existing and transcendent. A very real "pantheism" if you will (in the etymological sense of the word), but absolutely legitimate, since ultimately, if the reflective centers of the world are really "one with God", this state is not obtained by identification (God becoming all), but by the differentiating and communicating action of love (God all in all) – and this fundamentally orthodox and christian. (*The Human Phenomenon*, 1938–40, I, 223 E; 344 F)

PANTHEISM, ANCIENT

Synonym of pantheism of diffusion.

> In this active participation of our beings in a collective task (a task whose reality is visible at the end of every scientific avenue) the nebula of ancient pantheisms condenses and takes shape at the heart of the modern world. Instinctive, sentimental and passive acceptance of the cosmic powers is succeeded, in the human, by the rational devotion and reflective collaboration of the element in a common task and ideal. (*Human Energy*, 1937, VI, 158 E; 195–6 F)

PANTHEISM, CHRISTIAN *see also* **Pan-Christism; Panentheism**
Synonym of pantheism of union.

> In fact, only a "pantheism" of love or a christian "pantheism" (where human beings find themselves super-personalized, super-centered, by union with Christ, the divine super-center), only such a pantheism interprets correctly and wholly satisfactorily human religious aspirations whose dream is consciously to become finally lost in unity. (*Introduction to the Christian Life*, 1944, X, 171 E; 200 F)

PANTHEISM, FALSE
Synonym of pantheism of diffusion.

> This "pan-christism", we can clearly see, has nothing of the false pantheist. What makes up the usual defect of pantheism is that, by placing the universal center below consciousness and below the monads, it is obliged to conceive "omega" as a center of mental dissociation, of fusion, of unconsciousness, of minimal effort . . . The danger of false pantheisms has disappeared. (*My Universe*, 1924, IX, 59 E; 87 F)

PANTHEISM, HUMANIST *see also* **Neo-pantheism, Humanist; Terrenism; Unification, Pantheism of**

> The humanist pantheisms around us represent a wholly youthful form of religion. A religion little or even uncodified (outside marxism). A religion without an apparent God, without any revelation. But a religion in a real sense, if we understand by this word a contagious faith to which you can give your life. Despite great differences of detail, a rapidly growing number of our contemporaries already agrees on recognizing the supreme value of existence in devoting oneself body and soul to a universal progress that expresses itself through the tangible developments of humanity. (*How I Believe*, 1934, X, 123 E; 143 F)

PANTHEISM, IDEALISTIC *see also* **Diffusion, Pantheism of**
Hindu concept according to which the self, far from influencing or enriching the individual determinisms of the world, is identified with a highly spiritual absolute.

> There are highly spiritualized forms of idealistic pantheism. Even with these, the unique substance is no more than a refinement of the astral matter of the theosophists or the ether of the physicists. Pantheists see the universal being as a basis likely to be revealed in the interplay of shifting forms but strictly independent of the substance of these forms. The absolute is an amorpha from which everything emerges and is swamped again. Individual determinations have no absolute value. (*Note on the Universal Element*, 1918, XII, 272–3 E; 389–90 F)

PANTHEISM, MATERIALIST *see also* **Diffusion, Pantheism of**
Concept that holds that the common root of being is to be found in elementary matter or energy, variable in form but unchangeable in nature. Materialist pantheism and idealistic pantheism are both opposed to pantheism of diffusion.

> Gladly, "unbelievers" of our day bow before the "God of Energy". But it is impossible to stop at this somewhat vague stage of materialist pantheism. Under penalty of being less evolved than the terms animated by its own action, universal energy must be a thinking energy. (*The Spirit of the Earth*, 1931, VI, 45 E; 54–5 F)

PANTHEISM, NATURALIST *see also* **Pantheism, Pagan**

> I must try to group together and look at my various fundamental ideas on matter; first, its attractions, due to a certain symmetry with the divine (immensity, power, stability, "fontal being", the reservoir of all knowledge and all response . . .), leading to paganism and naturalist pantheism. (Journal, 4 February 1916, 27; see also Bruno de Solages OCarm, *Teilhard de Chardin*, 286–7)

PANTHEISM, OLD *see also* **Diffusion, Pantheism of**

> Looked at from a particular point of view, nature is a drug that tempts us with nirvana and all the old pantheism. (Letter, 27 July 1915, *The Making of a Mind*, 60; *Genèse d'une pensée*, 73)

PANTHEISM, PAGAN
Synonym of pantheism of diffusion.

> To try to go back to being lost in the great initial reservoir, sensed as eternal, immense, infinitely fertile: the God-below. This is hindu and pagan pantheism . . . fusion with matter. (Notes and Sketches, 9 March 1916, *Journal*, 53; see also Bruno de Solages OCarm, *Teilhard de Chardin*, 294)

PANTHEISM, POPULAR
Synonym of pantheism of diffusion.

> Popular pantheism has always accepted without question that, in order to conquer the plural, we must eliminate: in order to hear basic harmony, we must create silence. (*Reflections on Two Converse Forms of Spirit*, 1950, VII, 221 E; 230 F)

PANTHEISM, SOCIALIST *see also* **Pantheism, Humanist**

> Because there are socialist and bolshevik pantheisms, we immediately become suspect when we speak of the unification of beings and people. (Notes and Sketches, Cahier 7, 4 February 1920, Bruno de Solages OCarm, *Teilhard de Chardin*, 295)

PANTHEISM, SPIRITUAL *see also* **Unification, Pantheism of**
Concept that holds that a purely immanent God is born of an aggregation of monads that, by the same act, become divinized.

> Besides the materialist pantheisms (that seek the "universal element" in a plastic principle of the world) we find a class of spiritual pantheisms (that believe they can find this element in a (vital or intellectual) plasmatic principle of the universe). We can imagine, for example, a theory in which universal being would be conceived in the form of a soul of the world, in process of being formed from the sum of all individual souls, its particles. This theory only differs from the one I adopt here in this respect: the universal center of the world is considered here as wholly immanent and the monads that are aggregated in it are (or rather become) wholly divine in becoming attached to it. (*Note on the Universal Element*, 1918, XII, 273 E; 390 F)

PANTHEISM, TRUE *see also* **Pan-Christism**
Synonym of pantheism of union.

(a) The only way we have of responding to the obscure promptings of the cosmic sense in us is to push a laborious interpretation of the world and ourselves to its final limits. Union by differentiation and differentiation by union. This structural law that we recognized . . . in the stuff of the universe reappears here as the law of moral perfection and the sole definition of true pantheism. (*Sketch of a Personalistic Universe*, 1936, VI, 83 E; 103 F)

(b) In the totalized christian universe (in the "Pleroma" of St. Paul), in the final analysis, God does not remain alone but he is all in all ("*en pasi panta Theos*"). Unity in and through plurality. (*Introduction to the Christian Life*, 1944, X, 171 E; 199–200 F)

PARAMETER *see also* **Systematics**

1 In geometry, variable factor forming part of the equation of a curve.
2 In biology, by extension, criterion whose variations (rate of growth) permit the measurement of evolutionary progress at a given moment. Teilhard notes three types of biological evolution:

- evolution by dispersion;
- evolution by instrumental differentiation;
- evolution by greater consciousness.

> Up to now biology, in forming theories, has scarcely noticed, scarcely studied, the "evolution of consciousness", hardly fitted by its very scope to provide points of support for systematics. But in this undoubtedly lies the basic movement of which the two other types of evolution are no more than harmonics – and in it alone we at last possess an absolute parameter for development, not only of life on earth, but of the world. (*The Spirit of the Earth*, 1931, VI, 27 E; 34 F)

PARA-PANTHEISM *see also* **Extension, Pantheism of**
Theoretical form of "pluralist" pantheism in which the individual monad becomes depersonalized and successively identified with other monads (like an actor playing a series of different roles).

In a slightly different form we could say that the pantheist (or cosmic) sense tends to be expressed according to one or other of the following three formulae:

(1) to become all beings (an impossible and mistaken act): para-pantheism;
(2) to become all ("eastern" monism): pseudo-pantheism;
(3) to become one with all beings ("western" monism): eu-pantheism. (*Reflections on Two Converse Forms of Spirit*, 1950, VII, 223 E; 232 F)

PAROUSIA
Presence of Christ in glory at the end of time bringing together the final center, omega, who is the term of the phenomenal world, and Christ-Omega, who consummates the totality of creation in the completion of his Mystical Body.

Doubtless, at this point the Parousia will be realized in a creation carried to the climax of its capacity for union. The unique act of assimilation and synthesis that has continued since the beginning of time will finally be revealed and the Universal Christ will appear like a flash of lightning amid the storm clouds of a slowly concentrated world. (*Universalization and Union*, 1942, VII, 84 E; 113 F)

PASSIVITIES *see also* **Activities; Diminution, Passivities of**
Opposite of activities. Concept expressing human dependency on transcendent energies that humans must endure either to perish or to be integrated in them.

The passivities . . . form half of human existence. The expression means, quite simply, that what is not done by us, is, by definition, undergone. (*The Divine Milieu*, 1926–7, 1932, IV, 52 E; 72 F)

PATTERN *see also* **Figure**
In an evolutionary dimension, evolutionary structures that determine the contours of reality.

(a) But, this apart, is there not a most revealing correspondence between the figure ("pattern") of two opposing Omegas: one postulated by modern science and the other experienced by christian mysticism? (*The God of Evolution*, 1953, X, 242 E; 290 F)
(b) The overall pattern of flora and fauna represents . . . the pattern of movement. (*On Progress*, 1921, Unpublished)

PERSON

One first advantage that appears when we analyze and then construct the cosmos by means of the human person as our chosen element, is that its past immediately takes natural shape. (*Sketch of a Personalistic Universe*, 1936, VI, 55 E; 71 F)

PERSONAL

Everything in process of personalization. The human person that progresses. The universe that possesses a personalizing curve, that is, prepares the appearance of the human person.

We often talk of the person as if it represented a (quantitatively) reduced and a (qualitatively) attenuated form of total reality. This is exactly the opposite of what should be understood. The personal is the highest state in which we can grasp the stuff of the universe. In its mysterious atomicity, moreover, something unique and intransmissible is concentrated, grain by grain. (*The Salvation of Mankind*, 1936, IX, 136 E; 177 F)

PERSONAL, PRINCIPLE OF THE CONSERVATION OF THE *see also* **Personal-universal**

Law that expresses:

1 irreversibility of the ascent of spirit in the universe;
2 conservation of a certain quantum of energy through metamorphoses beginning with the impersonal and ending with the completed personal;
3 conservation of each elementary person once it has been constituted within the personal-universal.

A principle of universal value appears to emerge from our outer and inner experiences of the world that we would call the "Principle of the Conservation of the Personal".
(1) At a first stage, the law of the conservation of the personal states that the ascent of spirit in the universe is an irreversible phenomenon . . .
(2) At a second stage, the principle of the conservation of the personal suggests that in universal evolution a certain quantum of energy is held in an "impersonal" state and destined to find itself wholly transformed in the end into a "personal" state . . .
(3) At a third stage, the principle of the conservation of the personal signifies that each individual nucleus of personality, once formed, is for ever constituted as "itself" . . .

In a universe where spirit is considered at the same time as matter, the principle of the conservation of the personal appears as the most general and satisfactory expression of cosmic invariance first suspected and sought by physics on the side of the conservation of energy. (*Human Energy*, 1937, VI, 160–2 E; 198–200 F)

PERSONAL, RELIGION OF THE *see also* **Mysticism, Western**
Basic characteristic of christianity and western mysticism. Orientation of convergent cosmogenesis towards the unity of a christic personal center, final and transcendent, leading to the birth and development of personal human centers during the evolution of the universe with a view to a final gathering of created personal beings in the Mystical Body.

> Far from contradicting my profound tendencies towards pantheism, christianity, rightly understood, has never stopped, precisely because it is the savior of the personal, guiding, clarifying and above all confirming them by supplying a precise object and starting point for experimental verification. (*Sketch of a Personalistic Universe*, 1936, VI, 91 E; 112 F)

PERSONAL-UNIVERSAL *see also* **Universal-personal**
Like the universal-personal, the personal-universal represents the synthesis of the final center (that contains an unlimited power of unification) and the complex totality in which its radial focal point is found. Overcomes the contradiction between personal singularity and the abstract universal.

The specific point of departure of the personal-universal is the radiation of the person in the direction of the all, radiation that is limited and partly virtual in human beings but infinite and fully actualized in the supreme center.

> In analyzing . . . the formation of the personality, we have been led to recognize the properties of spirit-matter in the stuff of the universe. Now another no less paradoxical aspect of this same stuff appears, revealed as necessary for any "extension of the person" beyond itself: I mean the personal-universal. The most incommunicable and hence the most precious quality of each being is that which makes him one with all the rest. It is consequently by coinciding with all the rest that we shall find the center of ourselves. (*Sketch of a Personalistic Universe*, 1936, VI, 64–5 E; 82 F)

PERSONALISM *see also* **Futurism; Universalism**
One of the three teilhardian pillars of the future. Three elements of faith in the teilhardian human front of liberation.

(a) Futurism (by which we understand the existence of an unlimited sphere of perfection and discovery), universalism and personalism . . . the three unshakable axes on which our faith in human effort can and must depend with complete assurance. Futurism, universalism and personalism: the three pillars of the future. (*The Salvation of Mankind*, 1936, IX, 137 E; 178 F)

(b) Christianity is universalist . . . Christianity is supremely futurist . . . Christianity is specifically personalist . . . (*The Salvation of Mankind*, 1936, IX, 149–50 E; 191 F)

PERSONALITY *see also* **Individuality**
Spiritual center of reflection, freedom and love, that emerges across a particular threshold of evolution. The spiritual center is clearly distinguished, by openness to unitive inter-personal relationships and to hyper-centration on the divine person, from the individual, the bio-psychic center that makes up its infrastructure (but that is solely defined in terms, not of union, but of separation).

PERSONALIZATION *see also* **Personal**
Centration, decentration and super-centration. Completion of the person through:

1 harmonization with all other personal centers (unanimization);
2 fulfillment of the human person in the Mystical Body through the influx of the divine person.

> If we are to be fully ourselves and fully alive, we must:
> (1) be centered on ourselves;
> (2) be decentered on the "other";
> (3) be super-centered on one greater than ourselves . . .
> Not only physically, but intellectually and morally, we are only human if we develop ourselves. (*Reflections on Happiness*, 1943, XI, 117 E; 129–30 F)

PERSONALIZATION, PAIN OF *see also* **Differentiation, Pain of; Metamorphosis, Pain of; Plurality, Pain of**
Positive effort and suffering at the basis of:

1 increasingly greater control of plurality by each free and reflective center;
2 participation of each center (where each in its turn is considered an element of plurality) in a higher synthesis oriented towards a final personal center. Term contains three elements: pain of plurality, pain of differentiation, pain of metamorphosis.

> For at least three reasons a personalizing evolution is necessarily painful: it is the basis of plurality; it progresses by differentiation; it leads to metamorphosis. (*Sketch of a Personalistic Universe*, 1936, VI, 85 E; 105 F)

PHENOMENOLOGY
In teilhardian thought, evolutionary dialectics (divergence, convergence, emergence). Knowledge of evolutionary time and space as opposed to knowledge of being that makes up ontology. Synthesizing vision of the totality of phenomena and the totality of the (internal and external aspects of the) phenomenon, as opposed to science where each discipline specializes in a given area of the real. Limited to the exteriority of the phenomenon. Forbids any extrapolation. Construction of the importance

of the whole of the real based on the human as center and a general law of recurrence (that represents the order of successive linkages of phenomena). Teilhardian phenomenology is a phenomenology of nature (and should not be confused with the phenomenology of consciousness of Edmund Husserl (1859–1938) and others).

> Essentially, the thought of Fr. Teilhard de Chardin is expressed not in a metaphysics but in a sort of phenomenology. (*My Intellectual Position*, 1948, XIII, 143 E; 173 F)

PHENOMENON

Appearance in time and space. Demonstrates the dimensions of exteriority and interiority of spirit. Physical, biological, psychological and social facts to the extent they can be described. Teilhard speaks of various phenomena: christian, cosmic, human, mental, mystical, religious, social, spiritual, etc.

> To be properly understood, the book I present here must not be read a metaphysical work, still less as some kind of theological essay, but solely and exclusively as a scientific study. The very choice of title makes this clear. It is a study of nothing but the phenomenon; but also, the whole of the phenomenon. (*The Human Phenomenon*, 1938–40, I, 1 E; 21 F)

PHENOMENON, CHRISTIAN

Christianity as a perceptible event in space–time giving rise to structures that can be described and analyzed. Notion does not affect the supernatural aspect of revelation.

> By the "christian phenomenon" . . . I understand the experiential existence, within humanity, of a religious current characterized by the following group of properties: an intense vitality; an unusual "adaptability" that allows it, unlike other religions, to develop better, and mainly, in the actual zone of growth of the noosphere; finally, a remarkable similarity, in its dogmatic perspectives (convergence of the universe on a self-subsistent and super-personal God) with everything we have learnt from the study of the human phenomenon. (*My Fundamental Vision*, No. 22, 1948, XI, 189 E; 204 F)

PHENOMENON, COSMIC

> I am convinced that [the material and spiritual] viewpoints need to be united, and that they soon will be, in a kind of phenomenology or generalized physics, where the internal face of things as well as the external face of the world will be taken into account. Otherwise, it seems to me that it would be impossible to cover the totality of the cosmic phenomenon with a coherent explanation, as science must aim to do. (*The Human Phenomenon*, 1938–40, I, 22 E; 49–50 F)

PHENOMENON, HUMAN *see also* Phenomenology

The human as an object of phenomenology rather than a subject of reflection. The human species is characterized by:

1 the power of reflection (reflective thought);
2 the power of co-reflection (power of totalizing its capacity for reflection);
3 Power of anticipating a critical point of ultra-reflection appearing as an opening to the irreversible – the only means of retaining in human beings what Teilhard calls a zest for evolution.

(a) I say "human" and not "man" on purpose in order to stress, on the level of basic understanding, that what influences my vision of humanity is neither social concentration nor zoological species but the (quasi-physico-chemical) perception of a certain extreme state achieved in its thinking (what we might call its "uranium") by the stuff of the universe. (*The Stuff of the Universe*, 1953, VII, 376 E; 399 F)

(b) By the expression "human phenomenon" we understand here the experiential fact of the appearance, in our universe, of the power of reflection and thought. (*The Phenomenon of Man*, 1930, III, 161 E; 227 F)

PHENOMENON, MYSTICAL

Evidence of certain believers of an immediate presence of God in the human and, ultimately, of a union of the human with God.

> The definitive discovery of the spiritual phenomenon is linked to the analysis (that science will finally undertake one day) of the "mystical phenomenon", that is, of the love of God. (*The Phenomenon of Spirituality*, 1937, VI, 112 E; 139 F)

PHENOMENON, RELIGIOUS *see also* Unanimization

In its widest sense, manifestation in a humanity in process of unanimization of faith in the all that is perceived as both incomplete reality and ideal value.

> The religious phenomenon, taken as a whole, is nothing less than the reaction of the universe as such to collective consciousness and human action in the course of development . . . The religious phenomenon is but one aspect of "hominization". And, as such, it represents an irreversible cosmic magnitude. (*How I Believe*, 1934, X, 118–19 E; 139 F)

PHENOMENON, SOCIAL

> The human social phenomenon is simply the higher form assumed on earth by the involution of the cosmic stuff on itself. (*My Fundamental Vision*, No. 19, 1948, XI, 184–5 E; 200 F)

PHENOMENON, SPIRITUAL

Progressive rise of consciousness in the universe across the thresholds of contingency and discontinuity.

(a) Taken as a whole, in its temporal and spatial totality, life represents the term of a large-scale transformation in the course of which what we call "matter" (in the most comprehensive sense of the word) is inverted, folded in on itself, interiorized, the operation covering, as far as we are concerned, the entire history of the earth. The spiritual phenomenon is not, therefore, a sort of brief flash in the dark: it traces a gradual and systematic passage from the unconscious to the conscious and from the conscious to the self-conscious. It is a change of cosmic state. (*The Phenomenon of Spirituality*, 1937, VI, 96–7 E; 121 F)

(b) Ultimately, the spiritual phenomenon represents the assured and definitive appearance of a cosmic quantum of consciousness . . . a quantum of personality . . . Humans mark nothing less than the origin of a new era in the history of the earth. In them, for the first time in the area open to our experience, the universe has become through reflection conscious of itself. It has become personalized . . . The spiritual phenomenon has entered into a higher and decisive phase by becoming the human phenomenon. (*The Phenomenon of Spirituality*, 1937, VI, 100, 102 E; 125, 127–8 F)

(c) Either the spiritual phenomenon is an unintelligible accident (and thereby signals the death of action) or it absorbs everything and imposes fundamental conditions on the structure of the universe around us. And among these fundamental conditions is the conservation and growth of the personal . . . (Letter to Bruno de Solages OCarm, 10 April 1934, *Lettres intimes*, 269)

PHILOSOPHY

Teilhardian philosophy is a philosophy of nature, a philosophy of evolution and convergence (philosophy of union), a positive philosophy that unites theory and reality. Teilhardian philosophy is wholly anti-manichæan.

(a) The problem of knowledge in human thought is gradually tending to be coordinated with, if not subordinated to, the problem of action. For ancient philosophy, "being" was, above all, "knowing". For modern philosophy, "being" is coming to be synonymous with "growing" and "becoming". (*Action and Activation*, 1945, IX, 174 E; 221 F)

(b) Was it not St. Thomas who, comparing (what today we should call the fixist) perspective of Latin Fathers like St. Gregory to the "evolutionary" perspective of the Greek Fathers and St. Augustine, said of the latter, "Magis placet" (II Sent, d.12; g.I.a.I)? Let us be glad to strengthen our minds by contact with this great thought! (*What Should We Think of Transformism?*, III, 1930, 154 E; 217 F)
Magis placet = it is more pleasing

PHYLETICS

Study of phyla.

By and large, the whole scientific world agrees in admitting that the design thus formed is essentially made up of ramified and divergent segments. But, whether on the internal structure and progressive transformation of these phyla or, more important, on their interconnections and laws (if these exist) of their succession and mass distribution in the biosphere, our knowledge is still sporadic or rudimentary. Despite an enormous quantity of accumulated material and ideas in circulation, a phyletics worthy of the name has not yet been successively formulated, as it should be, as an extension of modern genetics. (*A Defense of Orthogenesis*, 1955, III, 268 E; 383–4 F)

PHYLETIZATION *see also* **Orthogenesis**

Process of formation of the phylum.

Observed in a sufficient number of cases and over a sufficient period of time, repeated speciation gives birth throughout the ages to general alignments: the effect, we would say, of phyletization – or, that comes to the same thing, of orthogenesis; this latter word meaning no more here than the appearance in time, within related species, of a statistically orientated distribution. (*Note on the Reality and Significance of Human Orthogenesis*, 1951, III, 249–50 E; 355 F)

PHYLOGENESIS *see also* **Phyletization; Phylum**

Appearance and development of the phylum.

From this point of view, according to which the formation of tribes, nations, empires and, finally, the modern state is only an extension (with the assistance of a number of supplementary factors) of the mechanism that produced the animal species, human history . . . is particularly open to the study of the laws of phylogenesis. (*Man's Place in Nature*, 1949, VIII, 86–7 E; 127 F)

PHYLUM

Evolutionary fascicle composed of an immense quantity of morphological unities. One of the twelve major subdivisions of the animal kingdom, e.g. phylum chordata (to which humans belong). Consists of one class or a number of similar classes. By extension, Teilhard applies the term to any recognizable group-system, e.g. the Church.

(a) The phylum. The living fascicle, lineage of lineages . . . The phylum is first of all a collective reality . . . Next, the phylum is something polymorphous and elastic . . . The phylum, finally, is a dynamic kind of reality . . . What defines the phylum in the first place is its "initial angle of divergence", that is, the particular direction in which it groups and evolves as it separates from neighboring forms. What defines it in the second place is its "initial section" . . . Finally, and in conclusion, what not only completes the definition of the phylum, but moreover puts it in a category with other natural units of the world, is "its power and specific law of autonomous development". (*The Human Phenomenon*, 1938–40, I, 69–70 E; 121–3 F)

(b) In reality, to say that the Church is infallible is simply to recognize that the christian group as a living organism contains in itself to a superior degree the obscure sense and potentialities that enable it to find through innumerable gropings its way to maturity and completion. In other words, this is simply another way of reaffirming that the Church represents a supremely living "phylum". This being said, to localize, as catholics do, the permanent organ of phyletic infallibility in the councils, or, by an even more advanced concentration of christian consciousness, in the Pope (formulating and expressing, not his own ideas, but the teaching of the Church) fully conforms with the great law of "cephalization" that governs all biological evolution. (*Introduction to the Christian Life*, 1944, X, 153 E; 181 F)

(c) This is why, the more we observe the present great movements of human thought, the more convinced we feel that it is around christianity (taken in its "phyletic" form, that is, its catholic form) that the main axis of hominization is becoming ever more closely defined. (*Christianity and Evolution*, 1945, X, 186 E; 216 F)

(d) I should be happy to see you . . . substituting for a metaphysics that is stifling us an ultra-physics in which matter and spirit would be englobed in one and the same coherent and homogeneous explanation of the world . . . Thought will explode or evaporate unless the universe, in response to hominization, becomes divinized in some way . . . Christianity is the only living phylum that retains a divine personality. (Letter to Christophe Gaudefroy, 11 October 1936, *Lettres inédites*, 110, 111–12)

PHYSICAL

Sometimes used in the sense of "very real" as opposed to "moral" or "juridical". Borrowed from the theology of the Greek Fathers and applied not only to physical realities but also to spiritual and supernatural realities. Synonym of organic.

Love does not establish a simply extrinsic relationship between living beings, whose term is a simple moral completion of the one by the other. Love, in us, is the conscious trace of action that creates and fuses us. It is a factor of physical organization, of physical construction. (*Creative Union*, 1917, XII, 171 E; 219 F)

PHYSICISM *see also* Logicalism; Physical

Opposite of logicalism. Conception of the real as an organic and physical totality that can be built up, not only on the level of the cosmic, but also, by analogy, on the levels of the human and the divine. All three levels are ultimately synthesized.

The second answer, also very natural, that we can give to the crucial question "What is the nature of the function that governs the form and order of appearance of successive living beings?" is the following: "Living beings are arranged in their various categories, they order one another in their successive appearances, under the influence of the factor that, in its immediate reality, is physical, organic and cosmic. The universe is made up in such a way that living beings, taken in order of secondary causes, gradually encourage one another as a biological condition". Thus the defenders of physicism. (*Note on the Essence of Transformism*, 1920, XIII, 110 E; 127 F)

PHYSICIST *see also* **Physicism**

Corresponds to the abstract notion of physicism.

I have . . . become convinced of the existence, among human beings, of two irreconcilable categories of mind: the physicists (the "mystics") and the juridics. For the former, only organically structured being is beautiful; hence Christ, sovereignly attractive, must radiate physically. For the latter, being becomes disturbing as soon as it hides something greater and less definable than our human social relationships (considered in the sense of being artificial). (*My Universe*, 1924, IX, 55 E; 83–4 F)

PHYSICO-MORAL *see also* **Personal-universal; Spirit-matter**

Human level of spirit-matter.

We have already been confronted with complex dimensions such as spirit-matter and the personal-universal. We are now led correlatively to fuse into a common dimension two apparently opposite characteristics of experience. There are no longer around us separate physical and moral domains. There is only the physico-moral. (*Sketch of a Personalistic Universe*, 1936, VI, 72 E; 90 F)

PHYSICS *see also* **Phenomenology; Ultra-physics**

Science of the real. By analogy with the philosophy of Aristotle and others, designates not only the science of inanimate bodies but also a phenomenology of the cosmos centered around the human phenomenon and based on the law of complexity-consciousness.

For a century and a half physics, preoccupied with analytical research, was dominated by the idea of a dissipation of energy and the disintegration of matter. Now called upon by biology to consider the effects of synthesis, it is beginning to perceive that, parallel with the phenomenon of corpuscular disintegration, the universe historically displays a second process, as generalized and as fundamental as the first: I mean a gradual concentration of physico-chemical elements in nuclei of increasing complexity, each succeeding stage of material concentration and differentiation being accompanied by a more advanced form of spontaneity and psychic energy. (*Reflections on Progress*, 1941, V, 78 E; 103 F)

PLAN *see also* **Planning**
Corresponds to planning.

If, on the one hand, it becomes scientifically admissible that the technico-psychic organization of the human group represents an authentic extension of zoological evolution and if, on the other, it is undeniable that this organization, taken in its most active and sensitive part (I mean in the domain of reflective research and invention), is an internally planned operation: then, we must certainly yield to the evidence. (*Note on the Reality and Significance of Human Orthogenesis*, 1951, III, 253–4 E; 360 F)

PLANET
The word "planet" has many meanings in Teilhard's vocabulary. It can equally mean "earth" (or "world") or "universe" (or "cosmos").

PLANETARITY *see also* **Planetization**
Constitution of an organic whole that is interdependent with the entire planet.

I think we can reduce to three the stages I had to go through in succession, between the ages of 30 and 50 . . . before I was to become fully conscious of the extraordinary cosmic wealth concentrated in the human phenomenon. The first stage introduced me to the notion of human planetarity (existence and contours of a noosphere). The second showed me more explicitly the critical transformation undergone by the cosmic stuff at the level of reflection. And the third led me to recognize, under the influence of psycho-physical convergence (or "planetization"), an accelerated drift of the noosphere towards ultra-human states. (*The Heart of Matter*, 1950, XIII, 29–30 E; 39 F)

PLANETARY
The earth and its spheres (from the barysphere to the noosphere) that influence one another in a single evolutionary movement.

On the hills of Berkeley, the frontiers vanish between laboratory and factory, between the atomic and the social, and also, I should say, between the local and the planetary. (*On Looking at a Cyclotron*, 1953, VII, 351 E; 370 F)

PLANETIZATION, HUMAN *see also* **Phylogenesis; Ultra-human**
Convergence of humanity upon itself. Process whereby the different races and civilizations of *Homo sapiens sapiens* become synthesized to form a linked organic whole in which different spiritual elements converge and the ultra-human develops.

We will understand nothing of the human being anthropologically, ethnically, socially and morally, and, again, can never make any valid prediction regarding the future states of the human until we have seen that "ramification" (insofar as it subsists) now works with the single purpose – and in higher forms – of agglomeration and convergence. The formation of verticils, selection and struggle for life are from now on merely secondary functions, subordinated in the human being to a work of cohesion. The enfolding on itself of a fascicle of virtual species around the surface of the earth. A totally new mode of phylogenesis. This is what I have called "human planetization". (*The Human Phenomenon*, 1938–40, I, 171 and n. E; 269 and n. F)

PLANETIZE *see also* **Planetization**

For human beings on earth, on the whole earth, to learn to love one another, it is not enough that they should know themselves to be members of one and the same thing, but in "planetizing" themselves they must be conscious of becoming, without confusion, one and the same person. (*Life and the Planets*, 1945, V, 120 E; 152–3)

PLANETS, PLURALITY OF THINKING *see also* **Plurality of Thinking Planets**

PLANNING *see also* **Plan**

Rationalization of the real on the human level when evolution becomes sufficiently co-reflective to allow human finality to subject evolution to premeditated plans that reduce the effects of chaos.

A certain exaggerated planning would be fixism
= we need (1) an open and (2) a fluid planning.
Distinguish between planning and "fascist totalitarianism".
A democracy can be planned . . . (function of the preserved personality). (*Notes and Sketches*, 8 December 1945, in Claude Cuénot, *Nouveau Lexique*, 171)

PLEIAD *see also* **Monad; Multiple; Multitude**

Plurality of monads.

According to this hypothesis, each more spiritual monad is formed by the organization of a pleiad of less spiritual monads, under a wholly new principle of union. (*The Names of Matter*, 1919, XIII, 226 E; 450 F)

PLEROMA

Bringing of the universe to maturity. The whole of creation in union with Christ. Final completion of the supernatural organism in which the substantial one and the created multiple (many) are united without confusion in a totality that is a sort of triumph and generalization of being – without adding anything to God. Synthesis of the created and the uncreated in the Mystical Body of Christ, the great (quantitative and qualitative) completion of the universe in God.

Thus will be constituted the organic complex of God and the world – the Pleroma – the mysterious reality that we cannot say is more beautiful than God himself (since God could do without the world) but that we cannot consider wholly gratuitous, wholly accessory, without making the creation incomprehensible, the Passion of Christ absurd and our effort uninteresting. (*My Universe*, 1924, IX, 85 E; 114 F)

PLEROMIZATION *see also* **Pleroma; Trinitization; Union, Metaphysics of**

Constitution, development and completion of the Pleroma. The crowning of christogenesis.

(a) Creation, incarnation, redemption . . . the three mysteries really become, in the new christology, no more than three faces of one and the same fundamental process, of a fourth mystery . . . to which we should give a name to distinguish it from the other three: the mystery of the creative union of the world or pleromization. (*Suggestions for a New Theology*, 1945, X, 183 E; 213 F)

(b) No creation without incarnational immersion. No incarnation without redemptive atonement. In a metaphysics of union, the three fundamental "mysteries" of christianity appear as no more than three aspects of the same mystery of mysteries, of the pleromization (or unifying reduction of the multiple). (*My Fundamental Vision*, No. 31, 1948, XI, 198 E; 213 F)

PLEROMIZE *see also* **Pleromization**

God is entirely self-sufficient yet the universe brings him something vitally necessary: these are the two apparently contradictory conditions that participated being must henceforth explicitly satisfy (in order to fulfil its double function of "activating" our will and "pleromizing" God). (*Christianity and Evolution*, 1945, X, 177 E; 208 F)

PLURAL *see also* **Multiple; Multitude**

In a system of convergent cosmogenesis, for God, to create is to unite. And to be united is to be immerged. But to be immerged (in the plural) is to be "corpusculized". (*The Energy of Evolution*, 1953, VII, 262–3 E; 271 F)

PLURALITY *see also* **Multiple; Multitude**

Plurality and unity: the single problem to which all physics, all philosophy and all religion fundamentally bring us back. (*Sketch of a Personalistic Universe*, 1936, VI, 57 E; 73 F)

PLURALITY OF THINKING PLANETS

Teilhard accepts the possibility, the probability even, of thinking life elsewhere in the universe.

(a) The eventuality of the plurality of thinking planets is no longer what you call "a possibility" . . . It has become a positive probability (and even a very real probability)(something recently noted in passing by a leading American astronomer) . . . I am more than ever convinced that we shall need, sooner or later, a new Nicæa that will define the cosmic face of the incarnation. (Letter to Bruno de Solages OCarm, 16 February 1955, *Lettres intimes*, 459)

(b) Nothing proves that in accordance with some law of large numbers many obscure stars, many earths other than our own, may not already be scattered or may not still be expected among the galaxies . . . And if there have been, if there are and if there will be *n* earths in the universe, then what we earlier called "spheres", "isospheres" and "noospheres" no longer embraces the whole but applies only to an isolated element (a mega-corpuscle) of the total phenomenon. With centro-complexity no longer dealing solely with a single planet but with as many noospheres as there will be thinking planets in the heavens, the process of personalisation takes on a decisively cosmic aspect. Our minds can hardly take this on board. But the law of recurrence remains the same. And there will always be only one Omega. (*Centrology*, 1944, VII, 127 E; 133–4 F)

PLURALITY, PAIN OF *see also* Differentiation, Pain of; Metamorphosis, Pain of; Personalization, Pain of

Positive effort and suffering linked to the constantly renewed struggle against a plurality that never stops trying to tempt human beings at every stage of unification.

> Plurality (a residue of plurality inseparable from all unification in progress) is the most obvious source of our pain. Externally, it exposes us to jars and makes us sensitive to these jars. And internally, it makes us fragile and subject to countless kinds of physical disorders. Everything that has not "finished organizing" must inevitably suffer from its residual lack of organization and its possible disorganizations. Such is the state of human beings. (*Sketch of a Personalistic Universe*, 1936, VI, 85 E; 105–6 F)

POINT, CRITICAL *see also* Threshold

> Carried to a certain temperature or a certain pressure, bodies change state: they become liquefied or vaporized. Everywhere there are "critical" or singular points in the movement of matter. Why should there not be such points in the transformations of life? (*Man's Place in Nature*, 1932, III, 180–1 E; 353 F)

POLYGENISM *see also* Monogenism

Descent of humanity from several couples (plurality of mutants). Non-teilhardian biological concept borrowed from theology. Antonym of monogenism.

POLYPHYLETISM *see also* **Monophyletism**

Descent of humanity from several phyla. Biological concept rejected by Teilhard. Antonym of monophyletism.

> If the new perspectives of discontinuity and polyphyletism . . . should assume substance, the old evolutionary ideas of the nineteenth century, far from vanishing like a mirage, would on the contrary attain their true expression. (*The Transformist Paradox*, 1925, III, 100 E; 140 F)

POSITION, ANTHROPOCENTRISM OF *see also* **Movement, Neo-anthropocentrism of**

Antonym of neo-anthropocentrism of movement.

> (a) The old anthropocentrism was wrong in supposing human beings to be the geometrical and juridical center of a static universe. (*Man's Place in the Universe*, 1942, III, 228 E; 320 F)
>
> (b) Since Galileo it might seem that we had lost our privileged position in the universe. Under the increasing influence of the combined forces of invention and socialization, we are now in process of regaining our leadership: no longer in stability but in movement; no longer the center but the leading shoot of a world in growth. Neo-anthropocentrism, no longer of position, but of direction in evolution. (*Evolution of the Idea of Evolution*, 1950, III, 246–7 E; 349 F)

PRE-BIOSPHERE *see also* **Biosphere; Pre-life**

Layer of already highly-complex organic matter corresponding to the phase of molecular evolution preceding the biosphere.

> By the very mechanism of its birth, the film in which the inside of the earth is concentrating and deepening emerges for our eyes in the form of an organic whole where no element can any longer be separated from the elements surrounding it. A new indivisible has appeared at the heart of the great indivisible, the universe. Truly, a pre-biosphere. (*The Human Phenomenon*, 1938–40, I, 38 E; 73 F)

PRE-CENTRIC *see also* **Centricity, Fragmentary**

> This must have taken shape, by means of increasingly complex atomic chains, an initial series of more or less slack and confused iospheres, marking (by the gradual fusion of partial interiors) the progressive stages, not of the "a-centric", but of the "pre-centric" towards centricity. (*Centrology*, 1944, VII, 106 E; 111 F)

PRE-CONSCIOUSNESS

Evolutionary stage preparatory to the emergence of consciousness. Notion derived retrospectively from the emergence of consciousness or, more correctly, reflective consciousness that gives rise to a germ of immanence (or interiority) even at the most elementary levels of arrangement.

> It is somewhere at (or even below) the level of cellular structure that, with the segments of pre-consciousness finally uniting along a closed curve, the first closed nuclei (the first centered corpuscles) appeared in the world. (*Centrology*, 1944, VII, 106 E; 111–12 F)

PRE-EMERGENCE *see also* Emergence

Property of the final pole of personalization that, in order to be supremely consistent and unifying, can only be conceived as anterior and transcendent to the elements that it super-personalizes by uniting them.

> Of this new God-the-Evolver, appearing at the very heart of the old God-the-Worker, we must, of course, and in the first place, maintain at all costs (and of cosmic necessity) the primordial transcendence: for, if he had not pre-emerged from the world, how could he be its issue and its consummation ahead? (*The Energy of Evolution*, 1953, VII, 262 E; 271 F)

PRE-LIFE

Evolutionary stage preceding and preparing the emergence of life or matter as a spiritual and vital but not yet actualized power. Notion underlying the continuous transitions between structures of organic and inorganic matter without denying the discontinuous threshold of the emergence of life. First of the four sections (pre-life, life, thought, super-life) in *The Human Phenomenon.*

> Beyond the albumens and proteins but still very far from the cell . . . there are certain enormous corpuscles. From the external, chemical point of view, consideration of these new objects absorbs us. But have we given sufficient thought to the fact that, if these particles are hyper-complex, it is necessarily and correlatively because they are hyper-centered and consequently contain a germ of consciousness? Below life, then, there is pre-life. (*The Atomism of Spirit*, 1941, VII, 33 E; 30 F)

PROGRESS

Non-linear (non-continuous) but irreversible process that:

1 moves the universe through the totality of time;
2 comprises an increase of consciousness that constitutes simultaneously an increase in being.

> Let us consider . . . the two fundamental equations or equivalents established earlier:
> Progress = rise of consciousness;
> Rise of consciousness = effect of organization. (*Reflections on Progress*, 1941, V, 69 E; 93 F)

PSEUDO-COMPLEX *see also* **Aggregate; Eu-complex**
Antonym of eu-complex.

(a) What discloses and measures eu-complex (or organic) arrangements of matter, both as reagent and parameter, as opposed to the purely fortuitous or mechanical groupings (pseudo-complexes) arising among atoms or molecules, is their appearance and ascent as psychic properties. (*Does Mankind Move Biologically Upon Itself?*, 1949, V, 253 E; 328–9 F)

(b) The type of super-grouping, towards which the course of civilization is driving us, far from simply representing some material aggregates ("pseudo-complexes") where elementary freedoms are neutralized by the effects of large numbers or mechanized by geometric repetition, belongs on the contrary to the species of "eu-complex" or arrangement where arrangement, because it is, and to the extent that it is, a generator of consciousness, is ipso facto classified as biological in nature and value. (*Man's Place in Nature*, 1949, VIII, 101–2 E; 146–7 F)

PSEUDO-PANTHEISM *see also* **Pantheism; Para-pantheism**
Synonym of pantheism of diffusion.

In a slightly different form we could say that the pantheist (or cosmic) sense tends to be expressed according to one or other of the following three formulae:

(1) to become all beings (an impossible and mistaken act): para-pantheism;
(2) to become all ("eastern" monism): pseudo-pantheism;
(3) to become one with all beings ("western" monism): eu-pantheism. (*Reflections on Two Converse Forms of Spirit*, 1950, VII, 223 E; 232 F)

PSYCHE, PSYCHIC *see also* **Centration; Consciousness; Interiorization; Psyche**

If the universe appears to be sidereally in process of spatial expansion (from the infinitely small to the immense), then in the same way, and even more clearly, it appears to be physico-chemically in process of organic enfolding (from the very simple to the extremely complicated) – this particular enfolding of "complexity" being linked experimentally to a corresponding increase of interiority, that is, of psyche or consciousness. (*The Human Phenomenon*, 1938–40, I, 216–17 E; 334 F)

PSYCHODYNAMICS *see also* **Energetics**
Aspect of energetics concerned with defining the means of conserving and promoting the human zest for action.

And, simultaneously, the necessity of tracing as quickly as possible the main lines of a spiritual energetics – or "psychodynamics" (as we speak of "thermodynamics") – devoted to the study of the conditions under that the human zest for auto-evolution and ultra-evolution, presently dissipated in a hundred different forms of faith and love, would be able to be grouped on itself, to be conserved and to be intensified. (*The Convergence of the Universe*, 1951, VII, 295–6 E; 308–9 F)

PSYCHOGENESIS *see also* **Noogenesis**

General process representing the origin and development of psychism through successive stages of evolutionary ascent.

> Geogenesis, as we have said, emigrating to biogenesis which ultimately turned out to be nothing else but psychogenesis. With and in the crisis of reflection, no less than the next term of the series is uncovered. Psychogenesis had led us to the human being. Now it vanishes, replaced or absorbed by a higher function: first to give birth to spirit, then later to all its developments – noogenesis. (*The Human Phenomenon*, 1938–40, I, 123 E; 200 F)

PSYCHOGENIC

Appearance and development of psychism through successive stages of increasing consciousness.

> In conformity with this rule, we must consider as possessing organic value in the living (whether we are concerned with a virus or with a human being) every arrangement – we shall call such a form of arrangement "psychogenic" – that results in a rise in the "psychic temperature" – or, if you prefer, an increase in interiority – of the grouping that has been arranged. (*The Convergence of the Universe*, 1951, VII, 288 E; 300 F)

PULL *see also* **Attract**

Vis ab ante (pull from above) or final cause acting through attraction towards a loving and lovable personal being. Antonym of push.

> Hominized evolution must henceforth include in its determinism, over and above the economic vis a tergo (or "push"), the "pull" of some powerful attraction, psychic in nature. (*The Reflection of Energy*, 1952, VII, 334 E; 349 F)

PURITY *see also* **Ethics; Impurity**

Unification of energies of the human soul in an evolutionary and ascending direction. Expression with both ethical value and ontological meaning.

> The specific act of purity . . . is to unite the inner powers of the soul in the act of a single, extraordinarily rich and intense passion. Finally, the pure soul is what, overcoming the multiple and the disorganizing attraction of things, tempers its unity (that is, matures its spirituality) to the passion of divine simplicity. (*The Struggle against the Multitude*, 1917, XII, 108 E; 146 F)

PUSH

Vis a tergo (push from below) or antecedent cause acting in an impersonal way according to necessity. Antonym of pull.

> I must first of all draw the attention of biologists to a remarkable feature of the system of auto-evolution that began . . . at the heart of the terrestrial noosphere, with the entry of human socialization into a compressive phase. I want to underline the gradual replacement of external pressure by internal attraction (of push by pull) as the "motive force" of evolution. (*The Phyletic Structure of the Human Group*, 1951, II, 169 E; 231–2 F)

QUANTUM *see also* **Metaphysics; Mysticism**

An amount of energy. Energy is presence (*The Christic*, 1955, XIII, 99 E; 114 F).

(a) The history of consciousness and its place in the world remain incomprehensible to anyone who has not seen beforehand that by the unassailable integrity of it as a whole, the cosmos in which we humans find ourselves engaged constitutes a system, a totum and a quantum: a system in its multiplicity – a totum in its unity – a quantum in its energy; all three, moreover, within a boundless contour. (*The Human Phenomenon*, 1938–40, I, 14 E; 38 F)

(b) At the present time it seems that the mass of humanity, influenced by new evolutionary views, is turning towards a morality of effort and union . . . linked to a metaphysics in which the universe is seen as a quantum of psychic energy flowing towards higher states of consciousness and spirituality. (*Can Moral Science dispense with a Metaphysical Foundation?*, 1945, XI, 133 E; 145–6 F)

(c) You are right, it is becoming more and more necessary for all of us to know the earth – as we know our own bodies. This is not a picture of tourism or the picturesque, but a question of "conscientization". The human mass spread across the globe is not a cloud of dispersed individuals. I am convinced it represents a certain given "quantum" of spiritual energy whose global maturation is necessary to the completion of every human being. (Letter to Max Bégouën, 17 September 1926, *Cahiers Pierre Teilhard de Chardin*, N° 2, 19)

(d) Ultimately, the spiritual phenomenon represents the assured and definitive appearance of a cosmic quantum of consciousness . . . a quantum of personality . . . Humans mark nothing less than the origin of a new era in the history of the earth. In them, for the first time in the area open to our experience, the universe has become through reflection conscious of itself. It has become personalized . . . The spiritual phenomenon has entered into a higher and decisive phase by becoming the human phenomenon. (*The Phenomenon of Spirituality*, 1937, VI, 100, 102 E; 125, 127–8 F)

R

R
Symbol for reflection.

R1 *see* **Reflection, First**

R2 *see* **Reflection, Second**

R3 *see* **Reflection, Third**

RADIAL *see also* **Energy, Radial; Tangential**
Psychic or spiritual.

> Our experience makes it perfectly clear that, during the course of evolution, emergence only happens successively and in mechanical dependence on what precedes it . . . The radial as function of the tangential. The pyramid whose summit is supported by its base. (*The Human Phenomenon*, 1938–40, I, 192–3 E; 300–1 F)

RADIATION, MUTATION BY
Differentiation of living forms following a process of divergence operating in a number of limited directions that, instead of dispersing, progress in the direction determined by adaptation to different biotypes (environment or conditions of existence, air, soil, water, etc.), e.g. marsupials confined to Australia.

> By radiation . . . when new variations, instead of dispersing without going much further, follow a number of limited but progressive directions determined particularly by precise conditions of existence and environment. (*The Movements of Life*, 1928, III, 145 E; 204 F)

REALISTIC *see also* **Juridic**
Ontological level analogous to the physical and organic but deeper than the juridic to which it is opposed.

> From the realistic point of view that characterizes the whole of catholic christianity, the sacraments are not only a symbolic rite. They act biologically on what they represent in the domain of the life of personal union with God. Nowhere does this idea of the organic function of the sacrament appear more clearly than in the Eucharist (Mass and Communion). (*Introduction to the Christian Life*, 1944, X, 165 E; 193–4 F)

RECURRENCE *see also* **Recurrence, Law of**
Repetition of what appears to be an already manifested plan, combining at the same time a certain periodicity and something new.

> We know the atoms as sums of nuclei and electrons, molecules as sums of atoms and cells as sums of molecules. Could there not be, ahead of us, a humanity in formation, the sum of organized persons? And is not this, moreover, the only logical way of extending, by recurrence (in the direction of greater centered complexity and greater consciousness), the curve of universal moleculization? (*Man's Place in the Universe*, 1942, III, 229 E; 321 F)

RECURRENCE, LAW OF *see also* **Synthesis**
Universal cosmic law expressing coherence in the successive temporal phases of evolution through the reappearance at each stage of evolution of a repeated and totally renewed process. Between each successive process there is both similarity and dissimilarity, that is, analogy.

> The law of recurrence by which we are governed can be expressed as follows: a first multiple is followed by a first unification and at every successive stage of consciousness a new plurality is constituted to allow a higher synthesis. (*Sketch of a Personalistic Universe*, 1936, VI, 56 E; 73 F)

REDEMPTION *see also* **Creation**

REFLECTION (R) *see also* **Co-reflection; Reflection, Second; Reflection, Third; Ultra-reflection**
Faculty possessed by every human monad to become more perfectly centered on itself in order to become aware of its own faculty for freedom of action, for increasingly better adaptation. In terms of cybernetics, the specifically human power to use information to create information.

> Reflection: the state of consciousness become capable of seeing and foreseeing itself. To think is not only to know but to know that one knows. (*A Major Problem for Anthropology*, 1951, VII, 316 n.6 E; 330 n.3 F)

REFLECTION, FIRST (R1) *see also* **Reflection**
Individual human reflection.

REFLECTION, SECOND (R2) *see also* **Co-reflection**
Collective human reflection.

> In this new essay I shall clearly have to insist as usual on the collective (social) extension of reflection ("second reflection") under the influence of human convergence (= the great event or physical motion that it is so important to grasp at this moment). But I should particularly like to go into the extraordinary phenomenon of "diaphany" by which, for a humanity in course of

> "second reflection", the universe tends to reveal its (or a) center of convergence – not as a center engendered by energy in process of reflecting on itself but a center forming the very generative principal (mover) of this reflection. The phenomenon of the "third reflection", in fact, by which "omega" is reflected on (or "revealed" to) a universe become (through the first and second reflections) capable of reflecting omega. (Letter, 16 September 1951, cited (in part) in Claude Cuénot, *Teilhard de Chardin*, 362 E; 441 F)

REFLECTION, STEP OF *see also* **Superlife**

Passage from non-reflective to reflective life, from the biosphere to the noosphere, marking a discontinuous emergence in a process of continuous centration.

> From the "step of reflection" onward we truly have entered into a new form of biology, which is characterized by the following properties, among others:
> (a) the decisive emergence, in the life of the individual, of factors of internal arrangement (invention) above factors of external arrangement (use of the play of chance);
> (b) the equally decisive appearance, among the elements, of true forces of mutual closeness or estrangement (sympathy and antipathy) that relay the pseudo-attractions and pseudo-repulsions of prelife, or even of lower life . . .
> (c) finally . . . the awakening, in the consciousness of each individual element, of the demand for "unlimited superlife". (*The Human Phenomenon*, 1938–40, I, 218–19 E; 337 F)

REFLECTION, THIRD (R3) *see also* **Ultra-reflection**

Reflection of omega on co-reflective human consciousness.

> (a) Is the "revelation" of Christ any different from his "discovery" by collective human thought? . . . (Need for the 3rd reflection). (Retreat, Purchase, 23 June 1952, in Claude Cuénot, *Nouveau Lexique*, 212)
> (b) R3 – third reflection = christification of God and the divine. (Retreat, Purchase, 29 June 1952, in Claude Cuénot, *Nouveau Lexique*, 181)

REFLECTIVE *see also* **Pan-reflective; Reflection**

Outcome of reflection.

RESEARCH *see also* **Ethics; Morality; Responsibility**

Reflective groping that first appears on earth with *Homo sapiens sapiens* but only really blossoms in the modern world in an increasingly collective form with the value of a moral imperative and a new faith.

> (a) Thought artificially perfecting the very organ of its thought and life rebounding from the collective effort of its reflection. Yes, the dream that obscurely nourishes human research has fundamentally been to succeed in mastering, beyond any atomic or molecular affinity, that fundamental energy

that all other energies merely serve: united together, to seize the tiller of the world by putting our hands on the driving force of evolution itself. (*The Human Phenomenon*, 1938–40, I, 177 E; 278 F)

(b) If research is the very law of being advancing towards ever greater consciousness, we must find in and around us in the world something to justify this effort . . . If the world is to be able to maintain its effort of reflective research, it must preserve the results progressively obtained by this research. (*Observations sur la signification et les conditions biologiques de la recherche*, 1939, Unpublished)

RESIGNATION

State of communion with God only reached at the peak of human effort in the struggle against pain and evil. Christian resignation is quite the opposite of surrender.

I can only unite myself to the will of God (as endured passively) when all my strength is spent, at the point where my activity, fully extended and straining towards betterment (understood in ordinary human terms), finds itself continually counter-weighted by forces tending to halt or overwhelm me. Unless I do everything I can to advance or resist, I shall not find myself at the required point – I shall not submit to God as much as I might have done or as much as he wishes. If, on the contrary, I persevere courageously, I shall rejoin God across evil, deeper down than evil, I shall draw close to him; and at that moment the optimism of my "communion in resignation" necessarily coincides (by definition) with the maximum of fidelity to the human task. (*The Divine Milieu*, 1926–7, 1932, IV, 73 E; 99–100 F)

RESPONSIBILITY *see also* Morality; Movement, Morality of; Obligation

Homo sapiens sapiens has a total responsibility that makes him accountable for the whole world. The success of history depends on the contingency of human options, e.g. human unity may be brought about either by force of conquest or by force of love (caritas, agape): the choice is a one of collective human responsibility.

(a) One of the most important aspects of hominization . . . is the accession of biological realities (or values) to the domain of moral realities (or values). With the human and in the human evolution has become reflectively conscious of itself . . . In a spiritually evolutionary perspective . . . the initial basis of obligation is the fact of being born and developing in function of a cosmic stream. We must act in a certain way because our individual destinies depend on a universal destiny. In origin, duty is nothing other than the reflection of the universe in the atom. (*The Spirit of the Earth*, 1931, VI, 29 E; 36 F)

(b) Responsibility shows itself to be co-original and co-extensive in its genesis with the totality of time and space . . . The evolution of responsibility is nothing other than a particular aspect of cosmogenesis. Or, more exactly, it is cosmogenesis itself observed and measured, not (as we usually do) by the degree of organic complexity or psychic tension, but by the degree of continually rising inter-influence within a multitude progressively concentrated on itself in a convergent milieu. (*The Evolution of Responsibility*, 1950, VII, 209 E; 215–16 F)

(c) Above all, human beings (deeply convinced of the truth of evolutionary ideas) see the greatness of their responsibilities extending almost to infinity before them . . . Suddenly, they discover in the depths of their being the fearful task of conserving, increasing and transmitting the fortunes of the world . . . For the briefest yet very real moment the success of the whole affair, of this immense universal childbirth, is in the hands of the least among us. (*Basis and Foundations of the Idea of Evolution*, 1926, III, 137 E; 191–2 F)

(d) Henceforth we must take account, more explicitly, of the obligations of human beings towards the collective and even towards the universe: political duties, social duties, international duties – cosmic duties (one might say) of the first order that include the law of work and the law of research. A new horizon of responsibilities is being opened up to our contemporaries . . . (*Note on the Presentation of the Gospel in a New Age*, 1919, XIII, 220 E; XII, 411 F)

REVERSAL

(a) The end of the world: the reversal of equilibrium, detaching the spirit, complete at last, from its material matrix, to rest from now on with its whole weight on God-Omega . . . The end of the world: critical point, simultaneously, of emergence and emersion, of maturation and escape . . . (*The Human Phenomenon*, 1938–40, I, 206 E; 320 F)

(b) In a convergent universe each element is completed, not directly by its own consummation, but by its incorporation within a higher pole of consciousness in which alone it can enter into contact with all the others. By a sort of reversal towards the other its growth culminates in giving and in excentration. (*The Grand Option*, 1939, V, 55–6 E; 76 F)

RICHNESS, UNITY OF *see also* West, Road of the

Mystical result obtained by the Road of the West.

Unity of abstraction and unity of richness: the respective dreams of the two mysticisms (of east and west). Western mysticism dreams of unity through the concentration of spirit and the world; it is the only mysticism capable of giving a new lease of life to the efforts we can make. (Lecture, 19 January 1933, notes by Jean Bousquet, in Claude Cuénot, *Nouveau Lexique*, 218)

S

SACRAMENT *see also* **Eucharist; Eucharistization**

Sacred symbol. Expresses objective reality. The Catholic and Orthodox Churches recognize seven sacraments. For Teilhard, the Eucharist is, in some way, "the first of the sacraments or, rather, the one sacrament to which all others are linked. And this for the good reason that the axis of the incarnation, that is, of creation, passes directly through the Eucharist" (*Introduction to the Christian Life*, 1944, X, 165–6 E; 194 F).

(a) From the realistic point of view that characterizes the whole of catholic christianity, the sacraments are not only a symbolic rite. They act biologically on what they represent in the domain of the life of personal union with God. Nowhere does this idea of the organic function of the sacrament appear more clearly than in the Eucharist (Mass and Communion). (*Introduction to the Christian Life*, 1944, X, 165 E; 193–4 F)

(b) After each communion this connection [between the christian and the Body of Christ] persists – even in an attenuated form – despite the disappearance of the sacred species that had temporarily raised it to a privileged degree of intimacy and growth. If this is how we see things, then sacramental communion, instead of being a discontinuous element in christian life becomes its frame. It is the accentuation and the renewal of a permanent state that binds us constantly to Christ. In sum, the whole life of the christian, on earth as in heaven, can be seen as a sort of perpetual eucharistic union. (*Note on the Physical Union between the Humanity of Christ and the Faithful*, 1920, X, 18 E; 24 F)

SCALE *see also* **Structure, Scaled**

When we first meet, at the beginning of the quaternary, the particular group of primates to which we belong, we can still clearly distinguish the ramified and divergent structure so characteristic of all other living groups that preceded it. The fossil humans of the Far East (*Pithecanthropus, Sinanthropus, Homo soloensis*) seem to form a truly independent or marginal "scale" that points to the existence elsewhere (in Asia and Africa) of other elements that are more central but still form other scales. (*My Fundamental Vision*, No. 12, 1948, XI, 176 E; 192 F)

SCALE, ZOOLOGICAL *see also* **Structure, Scaled**

Element in a series of developing living forms linked to one another and substituting for continuous linear developments a series of discontinuous pressures connected to each other in a single entity.

> In fact, taken together, these various indications cannot but suggest to our minds what I shall call the notion of "zoological scale"; a natural unity, I would say, subphyletic in order, defined by the following characteristics: a well-marked individuality (both in habitat and form), a low miscibility with other elements in the phylum, a considerable initial mutative power, an aptitude for extending itself at length in a residual form. This idea of "scales" and hence of a scaled structure in every phylum (and particularly in the human phylum) not only clarifies for us the physiognomy of the pithecanthropus group. It also has the advantage of providing us with a general method of division capable of deciphering the still confused mass of fossil men in a really natural and genetic order. (*Man's Place in Nature*, 1949, VIII, 67–8 E; 98 F)

SCHEMES, GENEALOGICAL *see also* **Speciation, Patterns of**

> Whether it is the oligocene cynodon, the pontian mustelidæ or the pliocene siphneidæ . . . after the "initial blank", there is always the same fascicle bristling with enveloping branches. And is not this just the same genealogical scheme that, in historical or proto-historical times, holds good for the birth and development of civilizations? (*The Phyletic Structure of the Human Group*, 1951, II, 148 E; 207 F)

SCHEMES, GENERAL *see also* **Archetypes, Structural**
Methods of knowledge that discover being (whereas structural archetypes depend directly on being).

> General schemes: "scaled structure"
> Dispersion – analysis
> Convergence – synthesis } union. (*Note on the Possible Preparation of a Lexicon*, 1951, in Claude Cuénot, *Nouveau Lexique*, 188–9)

SCIENCE

(a) Science alone cannot discover Christ. But Christ satisfies the yearnings that are born in our hearts in the school of science. (*Science and Christ*, 1921, IX, 36 E; 62 F)
(b) Science, in all probability, will be progressively more impregnated by mysticism (in order, not to be directed, but to be animated by it). (*My Universe*, 1924, IX, 83 E; 112 F)

SEGREGATION *see also* **Aggregation**
Geological, biological and (by extension) spiritual term:

1 on the scientific plane, process of separation;
2 on the ontological and religious plane, process whereby one or more

elements cease participating in the movement of union that constitutes the real and tend thereby to non-being. Antonym of aggregation.

In reality, the movement that aggregates the universe to Christ is a segregation. A portion of evil matter, definitely eliminated, will form the irreducible waste of the universal operation of salvation. (*The Names of Matter*, 1919, XIII, 235 E; 461 F)

SELF-

Prefix. Expresses the notion of the spontaneity of life and spirit, e.g. self-arrangement, self-centered, self-completion, self-complexification, self-conscious, self-consciousness, self-consistence, self-control, self-defense, self-determination, self-directing, self-evolution, self-evolver, self-knowledge, self-limiting, self-moving, self-operating, self-organization, self-orientation, self-perception, self-possession, self-subsistent, self-sufficiency, self-sufficient, self-transformation, self-ultra-hominization.

SELF-CONSCIOUSNESS

Consciousness of self.

We are neither more nor less than the portion of the Weltstoff that has emerged in self-consciousness. (*The Salvation of Mankind*, 1936, IX, 132 E; 173 F)

SELF-EVOLUTION *see also* **Auto-evolution**

Evolution of self.

Biological evolution, having arrived at a reflective stage ("self-evolution"), can only continue to function to the extent that it awakens in human beings clear prima facie evidence that the death-carrier can be broken. (*The Death Barrier and Co-Reflection*, 1955, VII, 403 E; 426 F)

SELF-SUBSISTENT *see also* **Self-sufficient**

That which has no need of anything other than the self to exist (applies particularly but not exclusively to God).

Of course, even if conquered (or, more exactly, "contoured"), thanks to the notion of radial energy, our scientific repugnance at admitting that reflection can, in becoming self-subsistent, escape any fall backwards, we remain face to face with another mystery. How, in fact, can we imagine, even in the most general way, the sidereal event of a noosphere reaching its superior critical point of co-reflection? What will happen on earth at that moment for an observer watching the spectacle from outside? (*The Singularities of the Human Species*, 1954, II, 266 E; 364 F)

SELF-SUFFICIENT *see also* **Self-subsistent**

That which has no need of the cosmos (of participated being) to exist (applies exclusively to God).

God is entirely self-sufficient yet the universe brings him something vitally necessary. (*Christianity and Evolution*, 1945, X, 177 E; 208 F)

SENSE

Purpose or direction. Intuitive spiritual quality.

(a) In recent times, as a result of the emergence in our internal vision of a universe finally contracted on itself and on ourselves through the immensity of time and space, it certainly seems the passionate awareness of a universal quasi-presence is tending to be aroused, corrected and generalized in human consciousness. A sense of evolution, a sense of space, a sense of the earth, a sense of the human . . . So many different and preliminary expressions of the same new need of unification. (*My Fundamental Vision*, No. 35, 1948, XI, 202 E; 217 F)

(b) To understand the world, knowledge is not enough. You must see it, touch it, live in its presence and drink the vital heat of existence in the very heart of reality. (*The Spiritual Power of Matter*, 1916, XIII, 71 E; 86 F)

SENSE, CHRISTIC *see also* **Christic; Sense, Cosmic**

Sense of the omnipresence of Christ whose mystical energy impregnates the universe.

Cosmic sense and christic sense: in me, two axes apparently independent of one another in origin; whose connection, convergence and, lastly, basic identity I only understand through and beyond the human, after much time and effort. (*The Heart of Matter*, 1950, XIII, 40 E; 51 F)

SENSE, COSMIC *see also* **Cosmic**

Sense of contact with the whole universe that enables the human to grasp its fundamental unity.

(a) Although still obscure or dormant in many people this cosmic sense of the one and the all appears to me to be both the most primitive and most progressive form of psychic energy into which the other energies of the world are gradually being transformed around us. (*Reflections on Two Converse Forms of Spirit*, 1950, VII, 218–19 E; 227 F)

(b) Cosmic sense is and can only be love. (*Sketch of a Personalistic Universe*, 1936, VI, 83 E; 104 F)

(c) The cosmic sense must have been born as soon as human beings found themselves facing the forest, the sea and the stars. And since then we find evidence of it in all our experience of the great and the unbounded: in art, in poetry and in religion. (*Sketch of a Personalistic Universe*, 1936, VI, 82 E; 102 F)

SENSE, HUMAN

Consciousness of humanity as a concrete and tangible totality, capable of controlling its own evolution and building its own future.

> We newcomers of the 20th century, we find ourselves coinciding with an event that, as amazing as the initial formation, vitalization or hominization of the earth, is developing at a pace proportionate to our experience: I would say the awakening of the human sense, that is, the consciousness of terrestrial thought that it forms an organized whole which, endowed with growth, is capable and responsible for some future. (*The Sense of Man*, 1929, XI, 13 E; 21 F)

SIN, ORIGINAL

Sin at its source, at the dawn of historic humanity. Theological expression.

1 In an individual sense, act by which human beings, in their first use of freedom, consider themselves complete (an end in themselves) and consequently deny God.

> Sin, in the will that commits it, is at first no more than a misguided and particularist attempt to attain the singularly desirable synthesis of being. Concupiscence tempts us with the bait of unity . . . what has been no more than a delicate stage in the synthesis of spirit suddenly becomes an acute and almost mortal crisis. (*The Struggle against the Multitude*, 1917, XII, 102, 104 E; 139, 142 F)

2 In a collective sense, act capable of immediate spatio-temporal extension to the whole of humanity, not only as a collection of individuals, but also as an organic whole, without reducing in any way responsibility for fault.

> Finally and above all, the theology of salvation appears perfectly respected and justified. Undoubtedly original sin, in this explanation, ceases to be an isolated act to become a state (affecting the human mass as a whole, as the result of a stream of sin punctuating humanity in the course of time). (*Reflections on Original Sin*, 1947, X, 196 E; 228 F)

3 In a cosmic sense, act seen as the counterpart of the progressive unification of the multiple while remaining, on a human level, an act of positive will.

> Original sin, taken in a general sense, is neither a specifically terrestrial malady nor an inherent feature of human generation. It simply symbolizes the inevitable chance of evil (*necesse est ut eveniant scandala* – scandals must necessarily arise) attached to the existence of all participated being. Wherever being *in fieri* (in process of becoming) is produced, suffering and sin immediately appear in its shadow, not only as a result of the tendency of creatures to inaction and egoism, but also (that is more disturbing) as the fatal accompaniment of their effort to progress. Original sin is the essential reaction of

the finite to the creative act. Inevitably, through the agency of all creation, it creeps into existence. It is the other face of creation. (*Fall, Redemption and Geocentrism*, 1920, X, 40 E; 53 F)

SOCIALIZATION *see also* **Compression, Socialization of; Expansion, Socialization of**

1 Threshold reached by every living form consisting of an association of individuals in a single species.

All the evidence shows that socialization (or association in symbiosis, subject to psychic connections, of historically free and strongly individualized corpuscles) is a primary and universal property of vitalized matter. All we need to do to be convinced is to note how (in proportion and according to the particular modalities of its "type of instinct") each animal line, having arrived at a specific level of maturity, reveals in its own way a tendency to group together in the form of supra-individual complexes a relatively large number of its constituent elements. On these pre-reflective levels (particularly with insects), the ray of socialization – however developed it may be – always remains weak, stopping short, for example, at the family group. (*Man's Place in Nature*, 1949, VIII, 79–80 E; 116 F)

2 Formation, actually in progress, of a human organic community of persons tending to bring them together in a new synthesis that unites them in a whole by constantly personalizing them further. Process distinct from economic socialization or socialism.

By human socialization, whose specific effect is to make the whole network of reflective scales and fibers of the earth bend back on itself, it is the very axis of the cosmic vortex of interiorization that pursues its course. (*The Human Phenomenon*, 1938–40, I, 220 E; 340 F)

SOCIODYNAMICS *see also* **Anthropodynamics; Energetics**

Branch of sociology concerned with discovering and determining the energetic conditions of the orientated process towards the ultra-human.

Note American efforts . . . to construct a "sociometry" (by mathematical research into the statistical constraints in the human phenomenon). This effort should be completed by the effort to establish a "sociodynamics" investigating the conditions of the "activation" of "human energy". (*The Phenomenon of Man*, 1954, XIII, 155–6 E; 200 F)

SOUL, COMMON HUMAN *see also* **Noosphere; World, Soul of the**

The organization of human energy, taken as a whole, is directed and pushes us towards the ultimate formation, over and above each personal element, of a common human soul. (*Human Energy*, 1937, VI, 137 E; 171 F)

SPACE–TIME *see also* **Space–time, Evil of**
Organization of space and time, on the biological level, in a single convergent whole where the dimension of space is integrated in the dimension of time (whose transverse section it represents). Not to be confused with einsteinian space–time (represented by the spatialization of time and the geometrization of matter).

> For every modern mind (to the extent that it is truly modern) consciousness is always apparent – a sense is born – in an absolutely specific universal movement by virtue of which the totality of things, from the highest to the lowest, is shifting together, as a whole, not only in space and time, but in a ("hyper-einsteinian") space–time whose particular curve makes everything that moves within it increasingly arranged. (*From Cosmos to Cosmogenesis*, 1951, VII, 256–7 E; 264–5 F)

SPACE–TIME SICKNESS *see also* **Complexity, Infinity of**
Pain caused by the infinitely large and the infinitely small but also by the infinitely complex (third infinity) in so far as its positive aspects have not been discovered.

> In its first stages, "space–time sickness" manifests itself most often by an impression of being crushed and useless faced with cosmic immensity. (*The Human Phenomenon*, 1938–40, I, 159 E; 252 F)

SPECIATION *see also* **Speciation, Patterns of**
Formation of species. Antonym of phyletization.

> The phenomenon of speciation (or formation of species) corresponds to the secondary appearance (by mutation) somewhere within a structured population of one or more statistical centers of morphological grouping. (*Note on the Reality and Significance of Human Orthogenesis*, 1951, III, 249 E; 354 F)

SPECIATION, PATTERNS OF *see also* **Orthogenesis**
Divergent drifts in series in process of active differentiation. Correlative with an orthogenetic axis because the patterns form statistical figures that end by developing along certain preferential axes of equilibrium, action and evolution. New statistical concept of orthogenesis.

> What we find here, in the case of human beings, is quite simply the general pattern of speciation identifiable in all other animal groups . . . Whatever the extraordinary characteristics of its "inflorescence" . . . the human group, taken sufficiently deeply, obeys the fundamental laws of speciation. Studied in its "stem", it reveals itself as a true phylum, endowed with an autonomous power of ramification and divergence: a complete phylum in which successive verticils of anatomically and geographically marginal forms are visible, probably framing . . . a specially adapted nucleus of more internal fibers. (*The Phyletic Structure of the Human Group*, 1951, II, 148–9 E; 206–7 F)

SPHERE *see also* **Circle**

Symbol. Corresponds to the circle. Structural totality giving rise to certain specific dominant qualities. The movement from one sphere to another implies a sufficiently important change of structure to allow a qualitatively different whole to appear. Hence, if complexification is continuous, the movement from one sphere to another supposes discontinuity.

(a) If any single perspective has been clearly brought out by the latest advances in physics, surely it is that in the unity of nature there are different types of spheres (or levels) for our experience, each one characterized by the dominance of certain factors that become imperceptible or insignificant in the neighboring sphere or level. (*The Human Phenomenon*, 1938–40, I, 23 E; 50 F)

(b) It is precisely because God is the center that he fills the whole sphere. (*The Divine Milieu*, 1926–7, 1932, IV, 102 E; 136 F)

SPIRAL *see also* **Creation, Evolutionary; Cycle; Enfolding**

Symbol of an evolutionary universe. Ascent illustrates the continuous process of evolutionary creation. Figures on medallion designed by Teilhard to express the idea that "everything that rises inevitably converges" (*Faith in Man*, 1947, V, 192 E; 242 F). Teilhardian evolution is always a spiral.

> It is easy for the pessimist to discount this period of extraordinary civilizations that collapsed one after the other. Is it not far more scientific to recognize once again, beneath these successive oscillations, the great spiral of life irreversibly rising in relays as it follows the main line of its evolution? Susa, Memphis and Athens may die. Yet an ever more organized consciousness of the universe is passed from hand to hand and grows more and more brilliant. (*The Human Phenomenon*, 1938–40, I, 146 E; 234 F)

SPIRIT

Principle of union. Higher state of the stuff of the universe. Power of synthesis and sublimation of the multiple. Irreversible direction of universal evolution. Transformation in the course of which matter interiorizes. In the world of phenomena, spirit does not appear in a pure state but in process of spiritualization. Spirit is one. Matter is multiple.

(a) By spirit I understand "the spirit of synthesis and sublimation" in which through constant trial and error is painfully concentrated the power of unity that is diffused in the universal multiple: spirit born within and as a function of matter. (*How I Believe*, 1934, X, 107–8 E; 128 F)

(b) Spirit is no longer the antipodes but the higher pole of matter in process of super-centration. (*The Atomism of Spirit*, 1941, VII, 56 E; 63 F)

SPIRIT, ATOMISM OF *see also* **Ethics; Morality**

Psychic aspect of the general tendency of the universe towards granulation resulting in the appearance of monads or "grains" of consciousness.

(a) The Atomism of Spirit. An attempt to understand the structure of the universe. (*The Atomism of Spirit*, 1941, VII, 21 E; 27 F)

(b) I have tried to demonstrate how the most traditional human morality takes on new form, new coherence and new urgency, how it is harmoniously integrated in the great body of cosmic energies in order to dominate them, once human beings, in order to regulate their behavior by going beyond the individualistic position of the "monad", resolutely adopts the point of view of the atom in order to judge and act. The idea . . . of a spiritualizing moleculization of matter does not only illuminate the stuff of the universe in its internal structure. The main lines of a whole new philosophy of life, a whole new ethics and a whole new mysticism are correlatively picked out by the same shaft of light. (*The Atomism of Spirit*, 1941, VII, 49 E; 56 F)

SPIRIT-MATTER *see also* **Entities, Paired; Phenomenon, Spiritual; Spirit; Spiritualization; Universe, Stuff of the**

Property of the stuff of the universe. There is neither matter nor spirit, only matter in process of spiritualization.

> There is neither spirit nor matter in the world: the "stuff of the universe" is spirit-matter. No other substance is capable of producing the human molecule. (*Sketch of a Personalistic Universe*, 1936, VI, 57–8 E; 74 F)

SPIRITUALITY, CREATION *see also* **Christ, Cosmic**

Expression only developed *expressis verbis* after Teilhard's death (1955). Emphasizes the inalienable goodness of creation, without denying the mysterious presence of evil: "original blessing" came before "original sin". Implicit in St. Irenæus (*c*.130–*c*.200), John Scotus Eriugena (*c*.810–77), St. Hildegard of Bingen (1098–1179), St. Francis of Assisi (*c*.1181–1226), Meister Eckhart (*c*.1260–1327) and, more recently, Teilhard de Chardin, it draws heavily on both eastern and western christian traditions. Its symbol is the Cosmic Christ.

SPIRITUALIZATION *see also* **Spirit; Spirit-matter**

1 Synonymous with spirit in a genetic and dynamic perspective where matter and spirit are seen as simply related variables.

2 Metamorphoses not by dissociation of matter through an ascesis of separation but by transformation of matter through the actualization of its latent spiritual power.

> Spirituality is not a recent accident, arbitrarily or fortuitously superimposed on the edifice of the world around us; it is a deeply rooted phenomenon whose traces we can follow backwards with certainty, until lost to view in

the wake of a movement that carries us forward . . . Hence our evidence that, from a purely scientific and experimental point of view, the real name of "spirit" is "spiritualization". (*The Phenomenon of Spirituality*, 1937, VI, 96 E; 120–1 F)

STAGES *see also* **Stages, Law of**

The theory allows us to envisage, as stages, intermediate critical points signifying around us the appearance of higher souls that will engulf but not destroy us. (*Sketch of a Personalistic Universe*, 1936, VI, 65 n.1 E; 83 n.1 F)

STAGES, LAW OF

Palaeontological law according to which evolution develops by replacing one form by another so that continuity embraces the discontinuity of successive emergences, e.g. successive waves of the human species.

We have pointed . . . to the fundamental law of stages by which all perceptible changes in life, instead of taking place in a continuous manner, operate in a series of successive waves that replace and pass beyond one another. (*The Movements of Life*, 1928, III, 148 E; 207 F)

STRUCTURE, FAN-LIKE *see also* **Structure, Scale-like; Structure, Scaly**

STRUCTURE, SCALED *see also* **Scale, Zoological; Structure, Scaly**

STRUCTURE, SCALE-LIKE *see also* **Structure, Scaly**

In the middle quaternary, despite the paucity of osteological documents dating from this period, the fan-like (or, more exactly, scale-like) structure of the human race is already very apparent. (*Paleontology and the Appearance of Man*, 1923, II, 50 E; 73 F)

STRUCTURE, SCALY *see also* **Scale, Zoological**

Metaphorical and explanatory framework of reference comparing races and species with the scales of a pine cone that drop off progressively revealing younger, greener scales beneath, the central core or axis being real yet unobservable. Particularly evident in the succession of human branches.

What is interesting to note at the moment is that the zoological species, even if they form . . . isolated scales, in every case cover and overlap like the leaves of conifers, in such a way as to construct (or at least to simulate) a stem, a tree, a bush if you like, in every case a regular and coherent whole. In a recent study, we tried to fix, apart from any transformist hypothesis, this scaly structure of phyla in the case of primates . . . In fact, it is extraordinary to see how easily, throughout the zoological domain, the overlapping or feathered structure of living beings continues from the smallest to the largest zoological group. (*The Transformist Paradox*, 1925, III, 85 E; 120–1 F)

SUFFERING *see also* **Passivities**
Apparent growth or diminution of being, capable of transformation by human beings, if they so wish, into spiritual energy. In a teilhardian perspective, evidence of a world in an incomplete and mutable state where the multiple is not yet unified.

> What a vast ocean of human suffering, the totality of suffering, is spread, at every moment, over the entire earth. But what is this mass formed of? Of blackness, lacunae, waste? Not, let us repeat, of potential energy. In suffering is hidden, with extreme intensity, the ascending force of the world. (*The Spirit of the Earth*, 1931, VI, 51 E; 65 F)

SUPER- *see also* **Supra-; Ultra-**
Prefix. Expresses the notion of greater value and fuller being, e.g. super-aggregation, super-arrangement, super-balance, super-biology, super-body, super-centration, super-center, super-centricity, super-cerebralize, super-charity, Super-Christ, super-communion, super-complex, super-complexification, super-condition, super-conscious, super-consciousness, super-create, super-creation, super-culture, super-differentiate, super-dimension, super-effect, super-ego, super-energetics, super-entity, super-equip, super-evolution, super-evolutionism, super-family, super-focal point, super-germinate, super-groping, super-hominization, super-human, super-humanization, superhumanity, super-idea, super-immanence, super-individual, super-individualize, super-interiorization, super-intimacy, super-involution, super-kosmos, superlife, super-live, super-love, super-matter, super-member, super-milieu, super-molecule, super-moleculization, super-nation, super-naturalize, super-naturalist, super-organ, super-organization, super-organism, super-personal, super-personalization, super-personality, super-physics, super-reality, super-reflective, super-science, super-sense, super-sexual, super-socialization, super-spiritualization, super-stretched, super-structure, super-transformism, super-union, super-vitalization . . .

SUPER-CENTRATION *see also* **Centration; Decentration**
Third stage of three-stage process of spiritual perfection (decentration, centration, super-centration). Super-centration is characterized by the discovery that the basis of human personal autonomy is found in participation in the life of a superior center.

> A center does not simply exist but builds itself up. This is what the facts tell us. Consequently, there is an indefinite number of disparate ways of matter

being centered. Following the axis of complexity, everything happens around us as though the stuff of the universe were distilling into a rising series of continually more perfect centers: this super-centration corresponds in physics to the accumulation in each nucleus of an ever greater number of more varied and better arranged particles; and this super-centration expresses itself, in psychology, in an increase of spontaneity and consciousness. (*The Atomism of Spirit*, 1941, VII, 31 E; 38 F)

SUPER-CENTER

Universal center of human and cosmic evolution.

The noosphere physically requires . . . the existence in the universe of a real pole of psychic convergence: a center different from all other centers that it "super-centers" by assimilating them; a person distinct from all other persons whom it fulfils by uniting them to itself. The world would not function if there did not exist somewhere ahead in time and space "a cosmic point Omega" of total synthesis. (*Human Energy*, 1937, VI, 145 E; 180 F)

SUPER-CHARITY

Evangelical charity whose dynamic and totalizing character is seen in modern human beings as a function of the increase of their vision.

(a) Super-charity . . . under the influence of the Super-Christ, our charity is universalized, dynamized and synthesized. (*Super-Humanity, Super-Christ, Super-Charity*, 1943, IX, 167 E; 212 F)

(b) The prefix "super" is used in these three expressions (super-humanity, Super-Christ, super-charity) to indicate not a difference of nature but a more advanced degree of realization or perception. (*Super-Humanity, Super-Christ, Super-Charity*, 1943, IX, 151 n.1 E; 193 n.1 F)

SUPER-CHRIST

Christ of the gospels whose cosmic dimension and evolutionary character is seen in a wholly new sense of urgency in modern human beings who have become more capable of penetrating his greatness.

By Super-Christ I most certainly do not mean another Christ, a second Christ different from and greater than the first; but I do mean the same Christ, the Christ of all time, seen by us, with an urgency and an area of contact, in an enlarged and renewed form and dimension. (*Super-Humanity, Super-Christ, Super-Charity*, 1943, IX, 164 E; 208 F)

SUPER-CONSCIOUSNESS

1 Higher state of collective consciousness reached by humanity when it arrives at the ultra-human stage.

A harmonized collectivity of consciousness, equivalent to a kind of super-consciousness. With the earth not only covered by myriads of grains of thought, but wrapped in a single thinking envelope until it functionally forms but a single vast grain of thought on the sidereal scale. The plurality of individual reflections being grouped and reinforced in a single unanimous act of reflection. (*The Human Phenomenon*, 1938–40, I, 178 E; 279 F)

2 Emergence of a personal-universal center seen not as the supernatural character of the parousia but as the natural term of research.

Truly an infallible rule since, by virtue of a curve inherent in the universe itself, we cannot follow it without drawing close (even in the darkest shadows) to some supreme and saving pole of super-consciousness. (*The Singularities of the Human Species*, 1954, II, 239 E; 332 F)

SUPER-CREATE

God cannot appear as prime mover (ahead) without becoming incarnate and atoning, without becoming christified before our eyes . . . Christ can no longer "justify" us without super-creating the entire universe by the same act. (*From Cosmos to Cosmogenesis*, 1951, VII, 263–4 E; 272 F)

SUPER-CREATION

Super-animation and transfiguration of the universe through the sanctifying grace of Christ whose redemptive incarnation continues creative action by intensifying it.

The history of the world traces an act of the absolute and the divine where the spiritualizing activity of beings emerges as a sacred energy. There can no longer be any question of simply opposing the one and the multiple, spirit and matter. We must seek and worship the one through the other. Divine unity overcomes the plural by super-creation, not by substitution. (*The Road of the West*, 1932, XI, 47–8 E; 54 F)

SUPER-EGO *see also* Ego

God as center of centers.

This accumulation of evidence that the extreme point of each of us (our ultra-ego we might say) coincides with some common term of evolution (with some common super-ego). (*How May We Conceive and Hope that Human Unanimization will be realized on Earth?*, 1950, V, 285 E; 372 F)

SUPER-HUMAN *see also* Super-humanity

Not a new being different to *Homo sapiens sapiens* but *Homo sapiens sapiens* itself carried to a higher level of potentiality in a fully formed noosphere.

Super-human = a collective (= a super-humanity) (a totalized élite). (Journal, 2 October 1945, in Claude Cuénot, *Nouveau Lexique*, 203)

SUPER-HUMANITY

Higher state that humanity appears destined to attain if it succeeds in becoming fully totalized on itself and submitting to the vivifying action of grace.

(a) By "super-humanity" I understand the higher biological state that humanity appears destined to attain if, carrying to the limit the movement from which it emerged historically, it succeeds, body and soul, in becoming fully totalized on itself. (*Super-Humanity, Super-Christ, Super-Charity*, 1943, IX, 152 E; 196–7 F)

(b) Is not a humanity grown capable of consciously taking its place in cosmic evolution and of pulsating as a whole (with its own wavelength, I should say) with a common emotion, is not such a humanity, whatever its residual imperfections and crises linked to its metamorphosis, already organically a true super-humanity in relation to the neolithic world? (*Human Energy*, 1937, VI, 123 E; 155 F)

SUPERLIFE *see also* **Reflection, Threshold of**

1 In a biological sense, the triumph of life over death on the elementary level of competing elements.

Who cannot see that to be effective the darwinian struggle for existence quite rightly presupposes among the competing elements an obstinate sense of conservation, of superlife, in which the very essence of the whole mystery meets and is concentrated? (*The Zest for Living*, 1950, VII, 233–4 E; 241–2 F)

2 In a religious sense, the triumph of life over death through access of the person to immortality.

From the "step of reflection" onward we truly have entered into a new form of biology, which is characterized by the following properties, among others: (a) the decisive emergence, in the life of the individual, of factors of internal arrangement (invention) above factors of external arrangement (use of the play of chance);
(b) the equally decisive appearance, among the elements, of true forces of mutual closeness or estrangement (sympathy and antipathy) that relay the pseudo-attractions and pseudo-repulsions of prelife, or even of lower life . . .
(c) finally . . . the awakening, in the consciousness of each individual element, of the demand for "unlimited superlife". (*The Human Phenomenon*, 1938–40, I, 218–19 E; 337 F)

3 In a teilhardian sense, access of the person to the collective conscious life that, by transcending individual conscious life, prepares the higher life engendered by the union of personal centers and persons in a hyper-personal focal point of love and irreversibility (begins at death but is only completed beyond death). Fourth of the four sections (pre-life, life, thought, superlife) in *The Human Phenomenon*.

(a) Once we have recognized and admitted its character and its dimensions of a "planetary synthesis", how (in harmony with the internal laws of the phenomenon) can we conceive of life ending on earth? By extinction or explosion, in death? Or, rather, by ultra-synthesis, in some superlife? (*The Atomism of Spirit*, 1941, VII, 41 E; 48 F)

(b) To create the flux that, with an increasing intensity, probably over hundreds of centuries yet to come, must draw us towards the above and the ahead, the repulsive (or negative) pole of a death to be avoided must of energetic necessity be matched by a second attractive (or positive) pole of a superlife to be attained: a pole capable of awakening and of always satisfying further, with time, the two characteristic demands of reflective activity – the need of irreversibility and the need of total unity. (*On Looking at a Cyclotron*, 1953, VII, 357 E; 376 F)

(c) Christianity can only survive (and super-live) . . . by sub-distinguishing in the "human nature" of the Word Incarnate between a "terrestrial nature" and a "cosmic nature". (Letter to Bruno de Solages OCarm, 2 January 1955, *Lettres intimes*, 450)

SUPER-MAN *see* **Super-human**

SUPER-MATTER *see also* *Omega*

Sort of neo-matter playing the role of the negative of omega, assuming without justification the attributes of the absolute.

> The discoveries made over the last hundred years with their unitary perspectives have brought a new and decisive impetus to our sense of the world, our sense of the earth and our sense of the human. Hence the surge of modern pantheisms. But that impetus will only end by plunging us back into supermatter if it does not lead to someone. (*The Human Phenomenon*, 1938–40, I, 190 E; 297 F)

SUPERNATURAL *see also* **Super-creation; Ultra-human**

Transphenomenal presence whose gratuitous action requires a natural maturation and completes through super-creation what natural forces cannot bring about.

> Nothing seems to me more feasible and fruitful (and consequently more imminent) than a synthesis between the above and the ahead in a becoming of a "christic" type in which access to the transcendent hyper-personal would appear conditional on the previous arrival of human consciousness at a critical point of collective reflection: the supernatural, consequently, would not preclude but rather require, as a necessary preparation, the complete maturation of the ultrahuman. (*Reflections on the Scientific Probability and the Religious Consequences of an Ultra-human*, 1951, VII, 279 E; 289–90 F)

SUPER-PERSONAL CENTER *see also* **Super-center**

Universal center of human and cosmic evolution.

A center different from all other centers that it "super-centers" by assimilating them; a person distinct from all other persons whom it fulfils by uniting them to itself . . . (*Human Energy*, 1937, VI, 145 E; 180 F)

SUPER-PERSONALIZATION *see also* **Collectivization**

In the case of human beings, collectivization or super-socialization can only mean super-personalization, that is, in the final analysis, sympathy and unanimity (since the forces of love alone possess the property of personalizing by unifying). (*Super-Humanity, Super-Christ, Super-Charity*, 1943, IX, 160 E; 204–5 F)

SUPER-SOCIALIZATION *see also* **Collectivization**

SUPER-SOUL

Area of spiritual expansion open to human psychism when, animated by faith in the rationality of the world, it transcends its present limits.

Either nature is closed to what we require of the future, in which case thought, the fruit of millions of years of effort, will suffocate, still-born, aborting on itself, in an absurd universe. Or else an opening does exist – of super-soul above our souls; but then, for us to commit ourselves to it, this way must open, with no restrictions, onto psychic spaces that nothing limits in a universe we can entrust ourselves to passionately. (*The Human Phenomenon*, 1938–40, I, 163 E; 258 F)

SUPRA- *see also* **Super-; Ultra-**

Prefix. Usually expresses transcendence but sometimes metamorphosis brought about by transcendent action within creation, e.g. supra-analogical, supra-cosmic, supra-evolutive, supra-freedom, supra-hominize, supra-human, supra-individual, supra-pansensorial, supra-personal, supra-personalization, supra-personalize, supra-phenomenal, supra-physical, supra-rational, supra-real, supra-structural, supra-substantial, supra-terrestrial . . .

SUPRA-EVOLUTIONARY

By the name "Point Omega" I have long meant and I understand again here an ultimate and self-subsistent pole of consciousness, sufficiently involved with the world to be able to gather to itself, through union, the cosmic elements carried to the limits of centration by technical arrangement yet capable through its supra-evolutionary (that is, transcendent) nature of escaping from the fatal regression that threatens (through its structure) every construction whose stuff exists in space and time. (*My Fundamental Vision*, No. 20, 1948, XI, 185 E; 200 F)

SUPRA-PERSONALIZE

> Union made us human by organizing, under the control of a thinking spirit, the confused powers of matter. It will now makes us "super-human" by making us into elements subject to some higher soul. Internal union has personalized us so far. Now external union is going to "supra-personalize" us. (*Sketch of a Personalistic Universe*, 1936, VI, 63 E; 80–1 F)

SURPASS *see also* **Detachment; Transcendence**

Transcend.

> Like the pagan, I worship a perceptible God. I even touch this God, through the whole surface and depths of the world of matter in which I am held. But, to take hold of him as I would wish (simply to continue to touch him), I must always go further, through and beyond all domination, without ever being able to rest in anything – borne at each moment by the creatures, and at each moment surpassing them, in a continuous welcome and a continuous detachment. (*The Mass on the World*, 1923, XIII, 125–6 E; 147–8 F)

SURVIVAL *see* **Superlife**

SYMBOL

Symbols possess a numinous quality, that is, a quality suffused with a feeling of the divine. In Teilhard, as in Carl Gustav Jung (1875–1961), they are the bridge between the divine and the human. Symbols and signs should never be confused. Signs have only one level of meaning (monovalent); symbols have many (multivalent). Signs are understood intellectually; symbols can only be felt intuitively.

> Jesus on the Cross is together symbol and reality of that immense labor over the centuries that has, little by little, raised up the created spirit in order to concentrate it in the depths of the divine milieu. He represents (and, in a real sense, is) creation that, sustained by God, reascends the slopes of being, sometimes clinging to things for support, sometimes tearing itself from them in order to go beyond them, and always compensating, by physical suffering, for its moral failures. The Cross is, therefore, not something inhuman but super-human . . . Christians are asked to live, not in the shadow of the Cross, but in the fire of its creative action. (*The Divine Milieu*, 1926–7, 1932, IV, 87–8 E; 118–19 F)

SYNERGY

> From the moment that geology reveals the first traces of the biosphere we can follow the extraordinary mingling of two matters, the inorganic and the organic – the latter perpetually infiltrating the former in order to alter its chemical cycles and conquer its physical layers by a continual synergy (since one dare not yet say symbiosis). (*Basis and Foundations of the Idea of Evolution*, 1926, III, 118 E; 167 F)

SYNTHESIS *see also* **Analysis; Recurrence, Law of**

Interior unity. Evolutionary process of the real. Consists of a progressive and unitive arrangement of elements in increasingly centered wholes. Leads to successively larger emergences, on the level of the real, and to a totalizing vision, on the level of knowledge, with cognitive progress corresponding to an ontological surplus, that is, an increase of being. The primacy of synthesis really introduces a new dimension of thought (for which the primacy of analysis remains a constant temptation).

> The law of recurrence by which we are governed can be expressed as follows: a first multiple is followed by a first unification and at every successive stage of consciousness a new plurality is constituted to allow a higher synthesis. (*Sketch of a Personalistic Universe*, 1936, VI, 56 E; 73 F)

SYNTHESIS, HUMAN

> There is nothing absurd about the idea of a concentration of humanity in the world – the idea of our continuing to be aggregated rather than disaggregated. Indeed, without it, I cannot see any interpretation that can be applied without contradiction to the totality of the human phenomenon. The hypothesis of a human synthesis in process is satisfactory because it is wholly coherent with itself and with the facts. (*Natural Human Units*, 1939, III, 211 E; 296 F)

SYSTEMATICS *see also* **Parameter**

Methodical classification.

(a) Apparently, there is no branch of the tree of sciences that is as old yet as modest as systematics . . . To know scientifically a thing (whether being or phenomenon) is to locate it in a physical system of temporal antecedents and spatial links. (*The Natural History of the World*, 1925, III, 103 E; 145–6 F)

(b) The enormous science of systematics has been developed over the last hundred years by a growing number of researchers in ever greater detail in constantly expanding domains. This science, whose initial purpose was to establish a simple logical or nominal classification of beings, has gradually become a veritable anatomy or histology of the layer of life on earth . . . Everything is classified: therefore, everything holds together. (*Basis and Foundations of the Idea of Evolution*, 1926, III, 121 E; 171–2 F)

SYSTEMS, NOOSPHERIC *see also* **Christ, Cosmic**

Teilhard accepts the distinct probability of a plurality of noospheres or "noospheric systems" in an unbelievably immense and organic universe: all systems, however, must converge on the same Point Omega.

(a) And if there have been, if there are and if there will be *n* earths in the universe, then what we earlier called "spheres", "isospheres" and "noospheres" no longer embraces the whole but applies only to an isolated element (a mega-corpuscle) of the total phenomenon. With centro-complexity no longer dealing solely with a single planet but with as many noospheres as there will be thinking planets in the heavens, the process of personalisation takes on a decisively cosmic aspect. Our minds can hardly take this on board. But the law of recurrence remains the same. And there will always be only one Omega. (*Centrology*, 1944, VII, 127 E; 134 F)

(b) Astronomy teaches us the higher unit of grouped matter in the direction of the immense is the galaxy. Similarly, biology tells us the higher, absolute unit in the direction of complexity is the reflective noosphere. Provided, of course, that through time and space "noospheric systems" do not happen, by chance, to be formed in the world: hypothesis which will appear less fantastic if we remember that, since life is under pressure everywhere around us, there is nothing to prevent the universe from producing (successively or even simultaneously) several thinking summits. (*Man's Place in Nature*, 1949, VIII, 114 n.1 E; 164 n.1 F)

T

TANGENTIAL *see also* **Energy, Tangential; Radial**
Physical.

> Megasynthesis in the tangential. And therefore, by this very fact, a leap forward of radial energies along the principal axis of evolution. Still more complexity: and therefore even more consciousness. (*The Human Phenomenon*, 1938–40, I, 172 E; 271 F)

TENSION, PANTHEISM OF *see also* **Détente, Pantheism; Union, Pantheism of**

TERRENISM *see also* **Marxism**
Religion of the earth. Faith in the world, faith in the human, faith in the future (the ahead). Denial of divine transcendence. Synonym for marxism in teilhardian thought.

> Communism should be more correctly called "terrenism". There is real seduction in this enthusiasm for the resources and future of the earth. (*The Salvation of Mankind*, 1936, IX, 139 E; 180–1 F)

THEOGENESIS
Two-phase process. In the first phase, God becomes trinitized. In the second, God is enfolded in participated being. Unlike christogenesis that expresses a coming in space–time, theogenesis is a largely analogical notion representing, not a "becoming" of God, but:

1. in God, the life and dynamism of the relationship between the three persons of the Trinity in whom God is united and trinitized (see also *My Fundamental Vision*, No. 28, 1948, XI, 194 E; 209 F);
2. in history, the creation and unification of the multiple as the preparation for christogenesis and the Pleroma that respond, like a necessary harmony, to the interior life of the Trinity.

(a) God enfolds himself in participated being by evolutive unification of the pure multiple ("positive non-being") born – in a state of absolute potentiality – through antithesis to pre-posited Trinitarian unity: creation. (*Christianity and Evolution*, 1945, X, 178, E; 209 F)

(b) I am definitely inclined to consider the creation as an organic continuation and completion of the Trinity . . .

(1) Trinity: cleavage of the one on itself, through an absolute position
(2) creation =

(a) projection of the shadow (non-being+, absolute multiple)
(b) reduction of the multiple to the one

Thus creation is a 2nd time in theogenesis, of a different order, but necessary to the plenitude of the process. (Journal, 3 January 1946, in Claude Cuénot, *Nouveau Lexique*, 206)

THEOLOGY

In general we can say that, if the dominant preoccupation of theology in the early centuries of the Church was to determine intellectually and mystically the relation of Christ to the Trinity, today it is vitally concerned with analyzing and defining the relations of existence and influence linking together Christ and the universe. (*Suggestions for a New Theology*, 1945, X, 176–7 E; 207 F)

THEOSPHERE *see also* Christosphere

Sphere of the divine. Divine milieu wholly completed at the end of time whose essence can be expressed through:

1 process of concentration of the universe in the final center;
2 process of irradiation of the final center to the entire universe.

Could not a later and final metamorphosis have been in progress since the christian birth of love: awareness of an "omega" at the heart of the noosphere; the movement of circles towards their common center; the appearance of the "theosphere"? (*Human Energy*, 1937, VI, 160 E; 198 F)

THINGS, INSIDE OF *see also* Things, Outside of

Psychic aspect of the stuff of the world, capable of infinite dilution or intense concentration, according to the hierarchy of the real. Teilhardian "key term".

(a) Since the stuff of the universe has an internal face at one point in itself, its structure is necessarily bifacial; that is, in every region of time and space, as well, for example, as being granular, co-extensive with its outside, everything has an inside. (*The Human Phenomenon*, 1938–40, I, 24 E; 52–3 F)

(b) At every degree of size and complexity, the cosmic corpuscles or grains are not only, as physics recognizes, centers of universal dynamic radiation but, somewhat like human beings, they possess and represent . . . a small "inside" (however diffuse or fragmentary it may be . . .) in which is reflected, in more or less rudimentary form, a particular representation of the world: psychic centers in relation to their own selves and, at the same time, infinitesimal psychic centers of the universe. (*Centrology*, 1944, VII, 101 E; 106–7 F)

THINGS, OUTSIDE OF *see also* **Things, Inside of**
Physical aspect of the stuff of the world, increasingly tenuous and differentiated. Inside and outside are a function of one another and represent, on the level of phenomena, the two faces of one and the same reality. Teilhardian "key term".

> The time has come for us to realize that to be satisfactory, any interpretation of the universe, even a positivistic one, must cover the inside as well as the outside of things – spirit as well as matter. (*The Human Phenomenon*, 1938–40, I, 6 E; 30 F)

THINGS, WITHIN OF *see also* **Things, Inside of**

THINGS, WITHOUT OF *see also* **Things, Outside of**

THOMAS AQUINAS, ST. *see also* **Fathers, Greek**
Tommaso d'Aquino (*c.*1225–74), theologian and doctor of the Church. Major influence on Teilhard. The basic teilhardian categories are becoming and process, those of thomism are being and substance.

> Within the Church we find St. Thomas Aquinas and St. Vincent de Paul side by side with St. John of the Cross. (Letter to Auguste Valensin SJ, 12 December 1919, *Lettres intimes*, 35)

THOUGHT *see also* **Reflection, Step of**
Third of the four sections (pre-life, life, thought, superlife) in *The Human Phenomenon.*

(a) After thousands of centuries of effort, terrestrial life, daughter of the cosmos, emerged into thought . . . (*The Spirit of the Earth*, 1931, VI, 27–8 E; 34 F)
(b) Everything that exists has a basis of thought, not of ether. (*Cosmic Life*, 1916, XII, 41 E; 48 F)
(c) Thought is an actual physical energy, sui generis, which succeeded over several hundred centuries in covering the entire face of the earth with a network of linked forces. (*Man's Place in Nature*, 1932, III, 180 E; 254 F)

THRESHOLD, CRITICAL *see also* **Emergence; Life, Breakthrough to; Reflection, Step of**
Point of evolutionary discontinuity marked by the appearance of new realities formed through:

1 the emergence of previously negligible factors (on a lower level);
2 the emergence of new factors.

Successive thresholds appear on the curve of evolution, from atomization passing through the threshold of reflection right up to the final threshold.

> Modern science has made us familiar with the idea of certain sudden and

radical changes appearing inevitable in the process of development, provided it can be carried far enough, always in the same direction. For a minimum modification in its arrangement (or in the conditions governing this arrangement) matter, having arrived at certain extreme levels of transformation, is open to a sudden change of properties or even of state. This notion of critical thresholds is currently accepted in physics, chemistry and genetics. Has not the moment come for us to make use of it to reconstruct the entire fabric of anthropology on a new and solid foundation? (*The Convergence of the Universe*, 1951, VII, 284–5 E; 297 F)

TIME, ORGANIC *see also* Space–time; Time, Cone of

Continuum in a convergent curve indissolubly linked to space in order to form the stuff of the universe or, more correctly, its organic structure so that each element of the real:

1. becomes fibrous instead of pointed, that is, indefinitely extended both forwards and backwards;
2. becomes one with the collection of fibers making up the all.

The perception of organic time that we are talking about here (time whose unrolling corresponds to the gradual, progressive and irreversible elaboration of a collection of organically linked elements), this new perception, we should say, does not offer, in itself, any explanation of things but only a more correct view of their qualitative integrity. (*Basis and Foundations of the Idea of Evolution*, 1926, III, 130 E; 183 F)

TIME, CONE OF

Symbol. Expresses the convergent acceleration of organic time using the image of the lines of a cone converging towards the summit.

What makes us so different and so demanding, compared to previous generations, is the awakening of our consciousness to a new setting of cosmic dimensions: the cone of time. In this particular milieu, interminably divergent behind us but positively convergent ahead, an unexpected connection can be seen . . . for the benefit of spirit, between totalization and personalization, between immanent evolution and creation. (*Super-Humanity, Super-Christ, Super-Charity*, 1943, IX, 151–2 E; 195–6 F)

TIMES, NEW

Evangelization of the new times . . . The particular apostolate I propose, that seeks to sanctify not only a nation or a social category but the very axis of the human drive towards spirit, contains two distinct phases: the one natural, providing an introduction to christian faith; the other supernatural, in which the (revealed) prolongations of terrestrial activity are developed. (*Note on the Presentation of the Gospel in a New Age*, 1919, XIII, 209, 214 E; XII, 395, 404 F)

TOTALIZATION

Energy pushing each center to enter into a wider synthesis.

> Let us imagine . . . human beings become conscious of their personal relations with a supreme personality, to whom they are led to add themselves by the entire play of cosmic activities. In such beings and starting with them, a process of unification has inevitably begun that will be divided into the following stages: totalization of each operation in respect of the individual; totalization of the individual in respect of itself; finally, totalization of individuals in collective humanity. All this "impossibility" takes place naturally under the influence of love. (*Human Energy*, 1937, VI, 147 E; 182–3 F)

TRADITIONALISM *see* Modernism

TRANS-

Prefix. Across, beyond, through. Expresses notion of qualitative metamorphosis following the crossing of a threshold, e.g. transcendence, transcendent, Trans-Christ, trans-christian, trans-christianity, trans-cosmic, trans-ego, trans-evolutive, trans-evolutionary, Trans-Exercises, trans-experimental, trans-figuration, trans-form, transformism, trans-formist, trans-hominization, trans-hominize, trans-human, trans-humanize, transience, trans-individual, trans-matter, transparence, transphenomenal, trans-planetary, transposition, trans-substantiate, trans-substantiation, trans-uranian . . .

TRANSCENDENCE *see also* Detachment; Surpass

1 Creation of a new synthesis where unity is the superior quality.
2 Order of things incompatible with the order of phenomena yet essentially linked to the latter through the creation, incarnation and redemption, in a gift of love and grace that overcomes the conflict between absolute being and participated being.

(a) Transcendent (that is, independent of evolution). (*Outline of a Dialectic of Spirit*, 1946, VII, 146 E; 152 F)

(b) Transcendent (reversal) is k.fu of omega. Everything holds together. Christian ascesis is simply an expression of a k. law. (k = cosmic, k.fu = cosmic function). (Retreat, Peiping, 27 October 1941)

(c) The future center of the cosmos, while still presenting us with the characteristics of a "limit", must be considered as having from the beginning, by something of its own, emerged in the absolute . . . In it everything ascends as though towards a focal point of immanence. But from it everything also descends as though from a summit of transcendence. (*Sketch of a Personalistic Universe*, 1936, VI, 70 E; 88 F)

TRANSFORMATION, CREATIVE

Synthesis of continuity and discontinuity, permanence and emergence. Process that produces a wholly new being from a preexistent being. Expresses an immanence of creative action in secondary causes.

> Most of the difficulties met by scholasticism in face of the historical evidence of evolution suggest that it neglects to account (in addition to creation and education) for a third sort of perfectly defined movement: creative transformation. Alongside "*creatio ex nihilo subjecti*" and "*transformatio ex potentia subjecti*", there is room for an act "*sui generis*" that, making use of a pre-existing created being, enlarges it into a wholly new being. (*On the Notion of Creative Transformation*, 1919, X, 22 E; 30–1 F)

Creatio ex nihilo subject = created with pre-existence of subjacent matter

Transformatio ex potentia subjecti = transformation by causing subjacent matter to act

TRANSFORMISM

Transformation of one living species into another.

1 In a general sense, application, in the case of life, of the law that conditions knowledge of all things capable of being perceived by the senses or the mind (science).

> Transformism is no longer a simple hypothesis. It is a general method of research, accepted in practice by all scientists. More widely still, it is only the extension to zoology and botany of a form of science (historical science) that increasingly governs all human sciences (physico-chemistry, religion, institutions, etc.). (*What Should We Think of Transformism?*, 1930, III, 152 E; 215 F)

2 In a specifically biological sense, recognition of the natural law of succession governing the appearance of living species.

> A natural grouping of animals in time and space is the assurance of living beings having come into the universe through a natural door; and a natural origin of living creatures is the guarantee of there being a natural explanation (that is, a scientific explanation) for the successive appearance of phenomena. But is transformism basically anything other than belief in a natural link between animal species? (*The Transformist Paradox*, 1925, III, 87 E; 123–4 F)

3 In an epistemological sense, condition governing all biological hypotheses.

> Broadly understood, as it should be, transformism is already no longer a hypothesis. It has become the form of thinking outside of which no scientific explanation is possible. (*The Transformist Paradox*, 1925, III, 87 E; 124 F)

TRANS-HOMINIZE *see also* **Ultra-hominize**

And we find we have effectively reached the heart of the difficulties caused by the problem of knowing whether, and how far, it is physically (planetarily) possible for human beings to become trans- or ultra-hominized. (*The Energy of Evolution*, 1953, VII, 369 E; 390 F)

TRANS-HUMAN *see also* **Ultra-human**

State reached by the ultra-human after crossing the threshold of the beyond, that is, after transcending space–time.

(a) Ahead of us, through the continually accelerating play of collective reflection, nothing less, beyond a wide fringe of the ultra-human, than the access to a final focal point in which the human, through concentration, succeeds in uniting with some trans-human. (*From Cosmos to Cosmogenesis*, 1951, VII, 262 E; 270 F)

(b) The psychic impossibility for our minds . . . of entering phenomenally into direct contact with any trans- or super-human thing or person whatsoever. (*A Phenomenon of Counter-Evolution in Human Biology*, 1949, VII, 186 E; 193 F)

(c) The critical point of planetary reflection, the fruit of socialization, far from being a mere flash in the night, corresponds on the contrary, to our passage, by involution or dematerialization, to another sphere of the universe: not an end of the ultrahuman, but its accession to some trans-human, at the very heart of things. (*From the Pre-human to the Ultra-human*, 1950, V, 296–7 E; 385 F)

(d) If we are to inflame human beings, we must henceforth put forward a God who is trans/ultra human (and not super/ultra human). (Retreat, Purchase, 19–26 August 1954, 5th day, in Claude Cuénot, *Nouveau Lexique*, 208; see also Jacques Laberge SJ, *Pierre Teilhard de Chardin et Ignace de Loyola*, 107)

TRANSIENCE

Transcendence. Antonym of immanence. Phenomenon of interaction determining relationships of exteriority between monads.

Complete exteriority or total “transience”, like absolute multiplicity, are synonyms of non-being. (*My Universe*, 1924, IX, 46–7 E; 75 F)

TRANSITION *see* **Transcendence**

TRANS-MATTER *see also* **Spirit; Spiritualization**

Synonym of spirit that, following an interiorization of matter, ends up being detached, with the superstructure no longer being dependent on the infrastructure.

At the most fundamental level, what presently affects our views on the mechanism (the ascesis) of spiritualization is that spirit has stopped being for us "anti-matter" or "extra-matter" to become "trans-matter". Spiritualization can no longer be effected in our eyes in disagreement or in discord with matter but in traversing and emerging from it. (*Note on the Notion of Christian Perfection*, 1942, XI, 105–6 E; 117 F)

TRANSPHENOMENAL

Ontological reality that is both opposed to the phenomenological and providing its base. Attainable through:

1 knowledge of being by dilating the collection of phenomena and by investigating the inside of things;
2 existential dialectics of action demanding a guarantee of the absolute and of immortality.

But it is just this supposedly impenetrable envelope of pure "phenomenon" that the rebounding trajectory of evolution pierces, at least at one point, since by nature it is irreversible. This does not mean we can see what lies beyond and behind that transphenomenal zone of which we now have an inkling . . . But at least we know something exists beyond the circle that restricts our view, something into which we shall eventually emerge. It is enough to ensure that we no longer feel imprisoned. (*The Human Rebound of Evolution*, 1947, V, 210 E; 269 F)

TRANSPOSITION

Transference. Translation.

(a) I am sometimes a little alarmed when I think about the transposition of the popular notions of creation, inspiration, miracle, original sin, resurrection, etc. that I must make if I am to be able to accept them. (Letter to Auguste Valensin SJ, 17 December 1922, *Lettres intimes*, 90)
(b) I am less concerned with working towards an "idealist transposition" of the universe than towards a "spiritual transposition". (Letter to Auguste Valensin SJ, 13 October 1933, *Lettres intimes*, 255)

TRAVERSE, TRAVERSING

Detachment, no longer by severing but by traversing and sublimating. Spiritualization, no longer by negation or by evasion from the multiple but by emergence. (*The Atomism of Spirit*, 1941, VII, 56 E; 63 F)

TRAVERSE, MYSTICISM OF THE

For some months now I have been trying to sketch the broad path followed by mysticism in seeking to solve the fundamental intellectual and spiritual problem: "How to explain and then surmount the multiple and arrive at unity?" It seems to me there are two theoretical solutions (both tried): the eastern solution ("one achieves unity by dissipating the illusion of the multiple through evasion or suppression") and the western solution – still hardly formulated, I should say ("one achieves unity by extending the powers – convergent by nature – of the multiple"). Mysticism of "detachment" or mysticism of the "traverse"? (Letter, 22 May 1932, in Claude Cuénot, *Teilhard de Chardin*, 141 E; 175 F)

TRAVERSING, DETACHMENT BY

Spiritual exercise to which Teilhard gives new direction. Depends on matter and nature transcending themselves in order to find in every thing and every being something bigger, something ahead of them, to love every thing and every being without becoming absorbed by them, to go beyond them by taking them with them and by transfiguring them.

(a) In itself, detachment by traversing is in perfect harmony with the idea of incarnation that encapsulates christianity. (*The Evolution of Chastity*, 1934, XI, 73 E; 79 F)

(b) Detachment, no longer by severing but by traversing and sublimating. Spiritualization, no longer by negation or by evasion from the multiple but by emergence. (*The Atomism of Spirit*, 1941, VII, 56 E; 63 F)

(c) From the broadly understood point of view of incarnation, detachment or renunciation becomes, above all, something found not so much in things but in every thing greater and ahead of itself – that enables us to love things without being absorbed by them, to go beyond them by taking them with us. (*Essai d'intégration de l'homme dans la nature*, Four conferences for the Marcel Légaut group, November-December 1930, in Claude Cuénot, *Nouveau Lexique*, 70)

TRAVERSING, RENUNCIATION BY *see also* Traversing, Detachment by

What the christian should no longer do is to withdraw, like the buddhist, from things by avoiding them but pass beyond by exploring them, measuring them, conquering them, to the full. To enjoy them for himself? Not at all. To extract and bring to God all the essence of beauty and spirituality they contain? Absolutely. Renunciation, again, but renunciation "by traversing" and creation where pain is simply the sign of the effort; not a renunciation by rupture, by reducing contact where suffering is wrongly given an absolute value. (*Christianity in the World*, 1933, IX, 107 E; 140 F)

TRIAL AND ERROR *see* Groping

TRINITIZATION

Analogical rather than genetic expression. Transcendent act whereby God both affirms himself as one and spreads, in his unicity, the richness of the three persons.

God himself, in a strictly true sense, only exists by uniting himself . . . The creative act assumes perfectly defined meaning and structure. Fruit, in some way, of a reflection of God, no longer in him, but outside him, pleromization (as St. Paul said), that is, completion of participated being through arrangement and totalization, appears in some way responsive or symmetrical to trinitization. (*My Fundamental Vision*, No. 27–8, 1948, XI, 194, 195 E; 209, 210 F)

TRINITIZE

In the very act by which his reality asserts itself, God . . . trinitizes himself. (*My Fundamental Vision*, No. 28, 1948, XI, 194 E; 209 F)

TRUTH

Reality to the extent that knowing and being are not separate. Structured totality ("coherence") that is not definitive but serves as a starting point ("fecundity") for a later synthesis that is subject to constant verification and correction.

Truth is nothing other than the total coherence of the universe in relation to each part of itself. Why suspect or underestimate this coherence because we ourselves are observers? There are those who constantly oppose what they call anthropocentric illusion to what they call objective reality. This distinction does not exist. The truth of human beings is the truth of the universe for human beings, that is, the truth, pure and simple. (*Sketch of a Personalistic Universe*, 1936, VI, 54–5 E; 71 F)

TURN, INWARD *see* Reversal

U

ULTRA- *see also* **Super-, Supra-**

Prefix. Beyond. Transformation of totality by emergence (by crossing a particular threshold), e.g. ultra-actual, ultra-arrangement, ultra-biological, ultra-centration, ultra-center, ultra-centric, ultra-cerebralize, ultra-christian, Ultra-Christ, ultra-christic, ultra-complex, ultra-complexification, ultra-condense, ultra-conscious, ultra-consciousness, ultra-corpusculization, ultra-cosmic, ultra-critical, ultra-development, ultra-differentiation, ultra-dimension, ultra-discover, ultra-divine, ultra-ego, ultra-evolve, ultra-evolution, ultra-hominization, ultra-hominize, ultra-human, ultra-humanization, ultra-humanize, ultra-living, ultra-material, ultra-materialism, ultra-mechanize, ultra-microscopic, ultra-molecule, ultra-monotheism, ultra-organic, ultra-organize, ultra-orthodox, ultra-personal, ultra-personalization, ultra-personalize, ultra-phenomenological, ultra-physical, ultra-physicize, ultra-reflection, ultra-reflective, ultra-responsibility, ultra-simple, ultra-socialization, ultra-source, ultra-specialization, ultra-specialize, ultra-speciation, ultra-synthesis, ultra-synthesize, ultra-synthetic, ultra-terrestrial, ultra-unify, ultra-vision, ultra-vivifying, ultra-Weltanschauung . . .

ULTRA-CHRISTIANITY *see also* **Neo-Christianity; Ultra-human**

Expression representing, neither a new religion nor a transcended christianity, but a later stage of evolution where christians will have become conscious of the plenitude of the incarnation in a humanity that will have crossed the threshold of the ultrahuman. Christianity of the future whose full dimensions will have been revealed.

> There is and always will be christianity but christianity reincarnated a second time (a christianity squared) in the spiritual energies of matter. Exactly the "ultra-christianity" we need at this moment to respond to the mounting demands of the "ultra-human". (*The Christic*, 1955, XIII, 96 E; 111 F)

ULTRA-CHRISTIC *see also* **Christic; God-above; God-ahead**

Future revelation of the total dimension of Christ and his energetic radiation as well as his synthetic importance as union of the God-above and the God-ahead.

> My constant action and vocation – to devote myself to making:

| the ultra-christic and
| the ultra-human
appear and develop through each other . . .
Christ can no longer maintain himself without revealing a cosmic value.
We need a Christ who is
| not only "super-naturalizing"
| but also ultra-humanizing. (Retreat, Les Moulins, 27 September 1950, in Claude Cuénot, *Nouveau Lexique*, 214)

ULTRA-DIFFERENTIATION *see also* **Differentiation**

"True union does not fuse: it differentiates and personalizes" . . . Unity at the base through dissolution or unity at the summit through ultra-differentiation. (*Reflections on Two Converse Forms of Spirit*, 1950, VII, 222 E; 230–1 F)

ULTRA-EGO *see also* **Ego**

We find ourselves solidly engaged in a process (cosmogenesis culminating in anthropogenesis) upon which our completion – or, one might say, our beatification – obscurely depends. Accumulating evidence suggests that the our goal (one could say our ultra-ego) coincides with some common term of evolution (with some common Super-ego). (*How May We Conceive and Hope that Human Unanimization will be realized on Earth?*, 1950, V, 285 E; 371–2 F)

ULTRA-EVOLUTION

The time now seems ripe for a small number of people representing the major living branches of modern scientific thought (physics, chemistry, biochemistry, sociology, psychology) to get together to focus their efforts on the following:

(1) to affirm and secure official recognition of the view that henceforth the question of human ultra-evolution (through collective reflection or convergence) must be studied scientifically.
(2) to research together the best means of verifying and dealing scientifically on every level with the problem and all its consequences.
(3) to provide the bases of a technology of ultra-evolution (both bio-physical and psychological) from the twofold point of view:
    (a) of conceivable planetary arrangements (e.g. in research and eugenics) for an ultra-arrangement of the noosphere;
    (b) of psychic energies to be generated or concentrated within the framework of a humanity in process of collective super-reflection on itself. This is the whole problem of maintaining and developing the psychic energy of self-evolution. (*A Major Problem for Anthropology*, 1951, VII, 317–18 E; 332 F)

ULTRA-HOMINIZATION *see also* **Ultra-human**
Passage to the ultra-human.

> Higher and clearly still incomplete, the thinking summit of the human: a still ascending summit, I must insist, because it is still in process of ultra-hominization. (*The Evolution of Responsibility*, 1950, VII, 208 E; 215 F)

ULTRA-HOMINIZE *see also* **Trans-hominize; Ultra-human**
Passage to the ultra-human.

(a) "Ultra-hominize" . . . simply expresses the idea of the human being extended beyond himself in a better organized form, more "adult" than the one with which we are familiar. (*Man's Place in Nature*, 1949, VIII, 109 n.1 E; 157 n.1 F)
(b) If we are to be ultra-hominized, we see the necessity for a "specific effort of convergence" that shows itself increasingly heavy to support: we have to bear the entire technico-mental organization of the earth. (*The Death Barrier and Co-Reflection*, 1955, VII, 399 E; 422 F)

ULTRA-HUMAN *see also* **Trans-human**
Future stage of evolution when a planetized and unanimized humanity will have transcended itself on the level of the senses (con-spiration, q.v.), thereby achieving unity of the noosphere and revealing its center (omega) ahead in the imminence of the parousia.

> Concerning the question of the developing reality of an ultra-human (that is, of the reality of an extension beyond the human of the process of evolution) . . . I should like to show here that a verifiable scientific reply is, in fact, already possible. (*A Major Problem for Anthropology*, 1951, VII, 313 E; 327 F)

ULTRA-LIVING
The spiritual to the extent that it is born phenomenally but constitutes a specific event taking place on the critical threshold of reflection and, consequently, of accession to an order that is higher than the living.

> It was no longer, as previously, towards some "ultra-material" but, on the contrary, towards some "ultra-living" that I was trying to identify and fix the ineffable ambience. (*The Heart of Matter*, 1950, XIII, 26 E; 34 F)

ULTRA-PERSONALIZATION
Future stage of evolution where the personal emerges across a critical threshold on a level higher than that of the human because noospheric union will further elevate and differentiate individual egos.

> The second condition . . . is that the irreversibility, thus revealed and accepted, must apply not to any one part, but to all that is deepest, most precious and most incommunicable in our consciousness. So that the process

of vitalization in which we are engaged may be defined at its upper limit (whether we envisage the totality of the system or the destiny of each separate element within it) in terms of "ultra-personalization". (*The Human Rebound of Evolution*, 1947, V, 206–7 E; 265 F)

ULTRA-PHYSICS *see also* **Phenomenology**

New discipline. Replaces metaphysics. Encompasses the inside and the outside of the real and forms the essentially evolutionary "proper" (or key) of the teilhardian vision. Starts phenomenally from physico-chemical matter and leads across the discontinuous thresholds of life and spirit to phenomenological meaning and, finally, ontological and spiritual sense.

(a) In the extra-temporal metaphysics of being this question could appear disrespectful. Before all creation, proclaims scholasticism, the absolute must exist in its fullness. For us who simply seek to construct a sort of ultra-physics combining the sum of our experience, the answer to the problem is not so positive. (*Sketch of a Personalistic Universe*, 1936, VI, 70 E; 88 F)

(b) I should be happy to see you . . . substituting for a metaphysics that is stifling us an ultra-physics in which matter and spirit would be englobed in one and the same coherent and homogeneous explanation of the world . . . Thought will explode or evaporate unless the universe, in response to hominization, becomes divinized in some way . . . Christianity is the only living phylum that retains a divine personality. (Letter to Christophe Gaudefroy, 11 October 1936, *Lettres inédites*, 110, 111–12)

(c) And this is why I am so bold as to offer here, in the form of a series of linked propositions, an essay in universal explanation . . . not an abstract metaphysics, but a realist ultra-physics of union. (*Centrology*, 1944, VII, 99 E; 105–6 F)

ULTRA-REFLECTION

Development beyond the present stage of noospheric consciousness through co-reflection unanimized and elevated by planetization, thus already centered on omega.

If we are to extrapolate into the future, the technical, social and mental convergence of humanity on itself compels us to allow for a paroxysm of co-reflection, at some finite distance ahead of us in time: a paroxysm that cannot be better (or otherwise) defined than a critical point of ultra-reflection. (*My Phenomenological View*, 1954, XI, 214 E; 235 F)

UNANIMIZATION *see also* **Con-spiration; Co-reflection**

Appearance of noospheric consciousness that:

1 unites center-to-center by further differentiating human persons in con-spiration and co-reflection to the extent that each forms a focal point of love and reflection.

2 synthesizes by differentiating (not by abolishing) the cultural contributions and spiritual currents of humanity that is unified and polarized towards research and adoration.

Everything leads us to suppose that the potentialities of the human as we know it have no other foreseeable limits but those of the evolutionary tension sustained in it by an ever finer sense of its organo-psychic unification; that is, in the final analysis, by the maintained and sharpened intensity of a field of unanimization. (*The Singularities of the Human Species*, 1954, II, 259 E; 356 F)

UNANIMIZE *see also* **Unanimization**

(a) In affording us a biological, "phyletic" outlet directed upwards, the shock that threatened to destroy us will have the effect of re-orientating, dynamizing and, finally (within certain limitations), of unanimizing us. (*Some Reflections on the Spiritual Repercussions of the Atom Bomb*, 1946, V, 147 E; 186 F)

(b) And how can this "essential" and "exponential" zest for unanimizing oneself be maintained to the end unless one bases it on the attraction ever more explicitly exercised on the species by the approaching center of its biological convergence? (*The Singularities of the Human Species*, 1954, II, 261 E; 359 F)

UNIFICATION, PANTHEISM OF *see also* **Convergence, Pantheism of; Identification, Pantheism of**

Synonyms: Pantheism, Humanist; Pantheism, Neo-humanist; Pantheism, Socialist; Pantheism, Spiritual. First movement of pantheism of convergence when the final center emerges from evolution on the level of co-reflection and so remains the effective culmination of natural forces. Concept understood by Teilhard as a stage destined to be transcended. Major classification of pantheism.

(a) Omega born of co-reflection remains an essential and ultimately simply collective pantheism of unification. (Conversation with F. Lafargue, July 1954, in Claude Cuénot, *Nouveau Lexique*, 147)

(b) Pantheism of identification, at the opposite pole of love: "God is all". And pantheism of unification, beyond love: "God all in all". Two isotopes of spirit. (*Reflections on Two Converse Forms of Spirit*, 1950, VII, 223 E; 231–2 F)

UNION *see also* **Centration; Differentiation; Identification; Union, Metaphysics of**

Phenomenal modality of divine creative action represented on all levels of the real by synthesis leading to fuller being. On the human level, union, far from leading to the fusion of egos in the all, differentiates them by exaltation and reciprocal enrichment. Opposite of identification and unification (that exclude a pre-existing focal point).

(a) True union (that is, spiritual union or union of synthesis) differentiates the elements it brings together. (*How I Believe*, 1934, X, 117 n.5 E; 137 n.1 F)
(b) Only union brought about in and through love has physically the property not only of differentiating but also of personalizing the elements it organizes. (*The Directions and Conditions of the Future*, 1948, V, 235 E; 302 F)
(c) Union with Christ supposes essentially that we make him the ultimate center of our existence – something that implies the radical sacrifice of egoism. (Letter to Auguste Valensin SJ, 12 December 1919, *Lettres intimes*, 31)

UNION, CREATIVE *see also* **Creation, Evolutionary**
Theory of cosmogenesis. Evolutionary creation. Process found at every stage of evolution: unification beginning with the limited multiple and continuing towards final unity through a hierarchy of increasingly higher centrations that cause the appearance of increasingly wider and better centered wholes. Supposes a pre-existing creative focal point that is only revealed in final unity on the level of phenomena.

Creative union is the theory that accepts, in the present evolutionary phase of the cosmos (the only phase known to us), that everything happens as though the one is formed by successive unifications of the multiple – and as though the more perfectly it centralizes under itself a larger multiple, the more perfect it becomes. (*My Universe*, 1924, IX, 45 E; 73 F)

UNION, DIALECTICS OF *see also* **Centrology; Union**
Process immanent to the real, derived from a transcendental focal point through which elements of the multiple become harmonized in order to become grouped around increasingly powerful synthesized centers. Movement of the real (equally a movement of the mind) that emphasizes synthesis rather than antithesis.

Christology: an essay on the dialectics of union. (*Centrology*, 1944, VII, 97 E; 103 F)

UNION, DIFFERENTIATED *see also* **Differentiation; Union**
Process that recognizes that the full potential of the individual person can only be realized within a greater whole that embraces us all – within a Person of persons who is identified in teilhardian thought with the Mystical Body of Christ. The more we become one in Christ, the more we become ourselves.

The universe is a continuous process of differentiation (“personalization”), the differentiation of elements being linked to their gradual unification (“differentiated union”). (Letter to Christophe Gaudefroy, 11 October 1936, *Lettres inédites*, 112)

UNION, METAPHYSICS OF *see also* **Metaphysics**
Metaphysical construction and phenomenological perspective according to which reality is to be found in a movement of unification and centration that differentiates infra-personal or personal unified elements.

> In a metaphysics of union . . . we see that once immanent divine unity has been achieved a degree of absolute unification is still possible, that would restore an "antipodial" aureole of pure multiplicity to the divine center. (*Christianity and Evolution*, 1945, X, 178 E; 208 F)

UNION, PANTHEISM OF *see also* **Convergence, Pantheism of**
Synonyms: Differentiation, Pantheism of; Eu-pantheism; Love, Pantheism of; Pantheism, Christian; Pantheism, True. Second movement of pantheism of convergence when the final center is recognized as pre-existent to evolution and revealed as a transcendent and supernatural focal point of union. Orthodox concept adopted by Teilhard. Major classification of pantheism.

> Omega born of the junction of co-reflection and a pre-existing (transcending) focal point of attraction . . . Pantheism of union under the influence of the third reflection. Omega is real. (Conversation with F. Lafargue, July 1954, in Claude Cuénot, *Nouveau Lexique*, 147–8)

UNIVERSAL-PERSONAL *see also* **Personal-universal**
Like the personal-universal, the universal-personal represents the synthesis of the final center (that contains an unlimited power of unification) and the complex totality in which its radial focal point is found. Overcomes the contradiction between personal singularity and the abstract universal. But its specific point of departure is the immensity of the all that returns to the personal center. Although their meanings are close, the second term of each composite expression can be considered as a qualification of the first.

> "Personalized" being, that makes us human, is the highest state in which we can apprehend the stuff of the world. Carried to its consummation, this substance must still contain to a supreme degree our most precious perfection. Thereby it can only be "super-conscious", that is, "super-personal". You may shy before the idea of a universal-personal. The association of the two concepts may seem monstrous. But this, I repeat, is spatial illusion. Instead of looking at the external, material sphere of the cosmos you should return to the point where the radii meet. (*How I Believe*, 1934, X, 114–15 E; 135 F)

UNIVERSALISM *see also* **Futurism; Personalism; Universalization**
One of the three teilhardian pillars of the future. Three elements of faith in the teilhardian human front of liberation.

(a) Futurism (by which we understand the existence of an unlimited sphere of perfection and discovery), universalism and personalism . . . the three unshakable axes on which our faith in human effort can and must depend with complete assurance. Futurism, universalism and personalism: the three pillars of the future. (*The Salvation of Mankind*, 1936, IX, 137 E; 178 F)

(b) True universalism claims to include without exception every initiative, every value, and every obscure potentiality in its syntheses. (*The Salvation of Mankind*, 1936, IX, 139 E; 180 F)

(c) The reason for the conflict between faith and progress that has done more harm to christianity than the most savage persecution seems to lie in a failure to come to terms with the three components of christian spirit (futurist, universalist and personalist). Christianity is universalist. But has it not continued to remain attached to a medieval cosmology instead of resolutely facing up to the temporal and spatial immensities to which the facts insist it must extend its views of the incarnation? Christianity is supremely futurist. But has not the very transcendence of the perspectives it maintains led it to allow itself to be regarded as extra-terrestrial (and so passive and soporific) instead of letting the sheer logic of its dogma make it supra-terrestrial (and so generator of maximum human effort)? Christianity is specifically personalist. But, again, has not the predominance given to the values of the soul encouraged it to present itself primarily as a legal and moral system instead of showing us the organic and cosmic splendors enclosed in its Universal Christ. (*The Salvation of Mankind*, 1936, IX, 149–50 E; 190–1 F)

UNIVERSALIZATION *see also* **Universalism**

Not a generalized abstraction. In convergent evolution:

1 better understanding of a concrete, progressive union of pre-human elements of the real;

2 better accomplishment and perception of the progressive union of reflective persons;

3 achievement, through successive stages, of the final center that synthesizes the totality of the multiple by differentiating the pre-human real and personalizing the multiplicity of human monads so that all opposition between universalization and personalization disappears.

Universalization of a human current, that is obtained by coercion that compels, by suppression that eliminates or by mechanization that de-humanizes, is complete; it reaches neither its maximum nor its equilibrium. It is only, if we reflect carefully, by an absence of universalism, either in the number of incorporated human elements or in the (insufficiently deep and total) form of contact achieved between them, that the democratic, communist and axis mystiques are still so violently opposed. Faithful to the end to the internal law of "greater universalization" that they all accept, the followers of these different movements must end up by discovering that, starting from different sides, they are assaulting the same mountain and they will inevitably meet on the same summit: "the personalizing grouping of the greatest possible number of men, in unanimity, through the heart". (*Universalization and Union*, 1942, VII, 93 E; 99–100 F)

UNIVERSALIZE *see also* **Universalization**

And here . . . my irrepressible urge to universalize whatever I love appears once again. (*The Heart of Matter*, 1950, XIII, 77 n.6 E; 38 n.1 F)

UNIVERSE

The word "universe" has many meanings in Teilhard's vocabulary. It can equally mean "cosmos" or "earth" (or "world" or "planet").

UNIVERSE, PERSONAL

Cosmos in process of laborious, personalizing unification that gradually integrates the "dust" of human souls (distinct from but dependent on God) in Christ through the build-up of humano-christian collective unity.

On the level of "simple life" all science teaches us is that union differentiates the elements it brings together. On the level of reflection, as we discover by self-observation, it personalizes us. By force of co-reflection, we must logically conclude, it totalizes itself into "something" in which all differences vanish on the borderline between universe and person. Such is the will of the law of complexity-consciousness, pushed to its limits. (*The Singularities of the Human Species*, 1954, II, 269–70 E; 367–9 F)

UNIVERSE, STUFF OF THE *see also* **Universal-Personal; Weltstoff**

Reality of the universe. Concrete being that forms the cosmos but that should not be confused with physical matter because it contains an "inside" or "within" (consciousness) and an "outside" or "without" (matter). Spirit-matter makes up the filament or web of the evolutionary universe whose unity is broken neither by moleculization nor by thresholds. Synonym of stuff of the world.

(a) Considered in its physical, concrete reality, the stuff of the universe cannot be split apart. But taken in its totality, as a kind of gigantic "atom", it forms the only real invisible there is (aside from thought where it centers and concentrates at the other end). (*The Human Phenomenon*, 1938–40, I, 14 E; 37–8 F)

(b) Since the stuff of the universe has an internal face at one point in itself, its structure is necessarily bifacial; that is, in every region of time and space, as well, for example, as being granular, co-extensive with its outside, everything has an inside. (*The Human Phenomenon*, 1938–40, I, 24 E; 52–3 F)

UPWARD *see also* **Above**

Distinguish the double movement of union:

(a) humanize (forward);
(b) Christify *stricto sensu* (upward) (Omega) = constant centro-christic omni-presence. (Retreat, Les Moulins, 30 August–7 September 1948, 5th day, in Claude Cuénot, *Nouveau Lexique*, 220–1)

VALORIZATION *see also* **Amorization**

Creative transformation. Innovation. Increasing awareness of the permanence of human achievement despite the constant rhythm of renewal. With amorization, one of the elements of moralization.

> Christians already no longer have any doubt they will have the last word. Because, in the end, only their "christic" vision of the world is capable of furnishing human effort with the two elements without which our action would be incapable of continuing its forward progress to the end:
>
> (1) valorization,
> (2) amorization . . .
>
> Evolution is the daughter of science. But, in the final analysis, it may well, perhaps, be faith in Christ that will preserve in us the zest for evolution. (*Catholicism and Science*, 1946, IX, 191 E; 241 F)

VERTICIL

Botanical term used metaphorically to describe the phyletic variations that produce secondary phyla corresponding to a fundamental variant or harmonic. Just as there are phyla of phyla, e.g. primates, with the same fan-like structures as an isolated phylum; equidæ, by extension, branches that diverge on a wider scale can also be described as "verticils". But in these two cases the notion of verticil implies a multiplicity of radiations on a relatively large layer.

(a) To understand the mechanism of this revival, we must continually return to the idea or symbol of groping. The formation of a verticil . . . is explained first of all by the phylum's necessity to pluralize itself in order to meet a variety of needs or possibilities. (*The Human Phenomenon*, 1938–40, I, 72–3 E; 126 F)

(b) Morphologically, like all other animal groups, the primates form as a whole a series of fans or overlapping verticils. (*The Human Phenomenon*, 1938–40, I, 103 E; 171 F)

VIA TERTIA *see also* **Traversing, Detachment by; Milieu, Divine**

Third way. Middle course between the God-above and the God-ahead, the via tertia leads to God through the earth, to spirit through matter, by traversing and sublimation. Expresses possible reconciliation through cosmogenesis between the abscissa, OX (below), and the ordinate, OY (above), through a via tertia, OR (ahead) (see also diagrams, *The Heart of the Problem*, 1949, V, 269 E; 349 F; *Two Principles and a Corollary*, 1948, XI, 160 E; 174 F).

(a) Previously, only two geometrical possibilities open to human beings were apparent: to love heaven or to love the earth. Now, in the new perspective of space, there is a third way: to reach heaven through the earth. There is a communion (true communion) with God through the world. (*The Awaited Word*, 1940, XI, 93 E; 111 F)

(b) Ever since human beings, in becoming human, began their quest for unity, they have never stopped oscillating, in their visions, in their ascesis or in their dreams, between a cult of the spirit that made them jettison matter or a cult of matter that made them deny spirit. Extenuation or degeneration. "Omegalization" makes us pass between the Scylla and the Charybis. Detachment, no longer by severing but by traversing and sublimating. Spiritualization, no longer by negation or by evasion from the multiple but by emergence. Such is the "via tertia" that opens up before us when spirit is no longer the antipodes but the higher pole of matter in process of super-centration. (*The Atomism of Spirit*, 1941, VII, 56 E; 63 F)

WAY, THIRD *see* **Via tertia**

WELTANSCHAUUNG
Worldview. Comprehensive and integrated view of the universe.

(a) These new perspectives on the unique nature of human beings . . . result so directly and so intimately from the whole modern scientific "Weltanschauung" that they are beginning, in fact, to color and permeate all that is conscious (or at least subconscious) in our own time. (*What the World is looking for from the Church of God*, 1952, X, 215 E; 256 F)

(b) It was to awaken myself to the sense of the creative, the additive and the hereditary in the common vision (Weltanschauung) slowly developed in the human spirit by all forms of research that I was finally confirmed in the perspective I now present. (*The Stuff of the Universe*, 1953, VII, 380 E; 402 F)

(c) "Metaphysics" should be understood as meaning any solution or vision of the world (of life) "as a whole" (any Weltanschauung) that is either imposed on the intelligence or categorically adhered to as an option or postulate. (*Can Moral Science dispense with a Metaphysical Foundation?*, 1945, XI, 130 E; 143 F)

WELTSTOFF *see also* **Things, Inside of; Things, Outside of; Universe, Stuff of**
Reality of the universe. Concrete being that forms the cosmos but that should not be confused with physical matter because it contains an "inside" or "within" (consciousness) and an "outside" or "without" (matter). Spirit-matter makes up the filament or web of the evolutionary universe whose unity is broken neither by moleculization nor by thresholds. Synonym of stuff of the universe and stuff of the world.

> But, in addition to this well-defined and well-established thermodynamic nucleus, should we not recognize the presence in the Weltstoff of certain structural elements that, negligible in physics and physical chemistry, assume a rapidly growing importance in the case of the extremely complex assemblies with which the sciences of life are concerned? (*The Energy of Evolution*, 1953, VII, 362 E; 382 F)

WEST, ROAD OF THE *see also* **Love, Pantheism of; East, Road of the**
Western mysticism that tends to present spirit as flowering over complexified matter with unity being obtained, not by suppression (unity of

détente), but by convergence of the multiple in a personal and personalizing focal point (unity of tension).

(a) There can be no conquering religion without mysticism. And no deep mysticism without faith in some unification of the universe. (*The Road of the West*, 1932, XI, 40 E; 47 F)

(b) The history of western mysticism could be described as a long attempt by christianity to recognize and separate, deep within itself, the eastern and western paths of spirituality: suppress or sublimate? Divinize by sublimation: it was on this side that the profound logic, the instinct of the nascent world, was to go. Divinize by suppression: it was in this direction that the accustomed ways of the ancient east were to push. (*The Road of the West*, 1932, XI, 51–2 E; 57 F)

WHOLE, THE

"Whole" and "All" are both translations of the Fr. "Tout". Teilhardian "keyword".

(a) Consideration of the "whole" imposes itself on every aspect of thought and dogma; the "whole" will be like a new dimension to which all our ideas must be readapted and in which they assume their true intelligibility. Then we could say that there is no "de indivuis" or "de singulis" philosophy, no concept holding strictly true except for the whole. (Letter to Auguste Valensin SJ, 20 October 1919, *Lettres intimes*, 18–19)

(b) Preoccupation with the whole has its roots in the most secret depths of our being . . . We can only think and, in the last analysis, can only dream of the whole. Perhaps we should add that sometimes the whole manifests itself directly to us, imposes itself almost intuitively on us. (*Pantheism and Christianity*, 1923, X, 57, 58 E; 74, 75 F)

(c) The universe only has complete reality in the movement that causes its elements to converge on some higher centers of cohesion, in other words, in the movement that spiritualizes the universe: nothing holds together except through the whole and the whole itself only holds together through its future completion. But, unlike free-thinking philosophers, christians can say they already find themselves in a personal relationship with the centre of the world: for christians this centre is Christ who really and non-metaphorically sustains the universe. (*Pantheism and Christianity*, 1923, X, 71 E; 74, 87–8 F)

(d) What do the people of our century need to compensate for and integrate in future progress the evil to which an insufficiently balanced perception of individual values will bring them? They must rediscover at all costs at the level of their current thought the sense and dominating passion for the whole. (*Basis and Foundations of the Idea of Evolution*, 1926, III, 140 E; 195 F)

(e) Hence we can say, in the final analysis (or rather "in the final synthesis") that in the end the person is for the whole and not the whole for the human person. But this because . . . the whole itself has become person. (*The Atomism of Spirit*, 1941, VII, 51 E; 58 F)

WHOLE, SENSE OF THE *see also* **Sense, Cosmic; Wholism**
Teilhard "was born with his sense for matter . . . since he was born with this sense, he called it natural" (Thomas King SJ, *Teilhard de Chardin*, 23).

> Human beings have always been sporadically attracted and intoxicated by the feeling of union with the whole in which they share. Recently, however, through the influence of improved understanding of the immensities and organicity of the universe around us, is not this elementary sense of the whole tending to become generalized in our consciousness – with this essential modification that the totality felt and desired no longer appears so much as an amorphous ocean in which we are dissolved but as a powerful focal point in which we are united, completed and concentrated? (*Psychological Conditions of the Unification of Man*, 1949, VII, 176 E; 182 F)

WHOLISM (HOLISM) *see also* **Whole, Sense of the**
Term first used *expressis verbis* by Jan Christiaan Smuts (1870–1950) in 1926. Teilhard raises it to the level of an evolutionary doctrine of universal application to express the fundamental unity of all things.

(a) It was the sense of consistence that led to the awakening and development of a dominant and victorious sense of the whole. (*The Heart of Matter*, 1950, XIII, 20 E; 28 F)

(b) Is it not obvious that christianity will only be able to breathe freely and spread its wings fully within the perspectives finally opened up to its spiritual potentialities by a true philosophy, not only of the whole, but of a convergent whole? (*Two Convergent Forms of Spirit*, VII, 1950, 226 E; 235 F)

WOMEN *see also* **Complementarity; Feminine**
Teilhard anticipated Vatican II and beyond on the place of women in the Church.

> It sometimes seems to me there are three weak stones sitting dangerously in the foundations of the modern Church: first, a government that excludes democracy; second, a priesthood that excludes and minimizes women; third, a revelation that excludes, for the future, prophecy. (Letter to Christophe Gaudefroy, 7 October 1929, *Lettres inédites*, 80)

WORLD
The word "world" has many meanings in Teilhard's vocabulary. It can equally mean "earth" (or "planet") or "universe" (or "cosmos").

WORLD, SACRAMENT OF THE *see also* **Eucharistization**
Process by which Christ, present and acting in the consecrated host, progressively assimilates humanity and, through humanity, the universe,

that is indispensable to the completion of the Mystical Body of Christ.

> It is beginning to be recognized, in the most conservative quarters, that, haloing the Eucharist, there is a communion with God through the earth – a sacrament of the world. (*The Evolution of Chastity*, 1934, XI, 73 E; 79 F)

WORLD, SOUL OF THE *see also* **Noosphere; Pantheism, Spiritual; Soul, Common human**

Platonic-stoic expression used by Teilhard in his early writings but later abandoned. Principle of unity that is immanent in the world that makes it both intelligible and lovable like a singular totality to which its constituent parts relate. The soul of the world is a "form", in the scholastic sense, that has been immanent in the universe from its beginnings. Not to be confused with the noosphere that is a phenomenal emergence that corresponds to hominization and, more precisely, to its phase of planetization.

(a) The soul of the world, whose life is drawn from the word infused in it, is at the same time the pivot necessary to the incarnation. It supplies ready to hand the matter intended to form the mystical body – it is what, dissimulated among innumerable creatures, envelops us in a living network, charged with grace and spirituality. (*The Soul of the World*, 1918, XII, 187 E; 255 F)

(b) If . . . the social unification of the earth is the state towards which evolution is really taking us, this transformation cannot contradict the clearest achievement of evolution in the course of history – the increase of consciousness and individual freedom. Like any other union, the collectivization of the earth, correctly carried out, must super-animate in us a common soul. (*Sketch of a Personalistic Universe*, 1936, VI, 80 E; 100 F)

WORLD, STUFF OF THE *see* **Universe, Stuff of; Weltstoff**

WORLDS, PLURALITY OF *see* **Plurality of Thinking Planets**

Select Bibliography

PRIMARY SOURCES

1 Collected Works in French

Dates shown below are the dates of publication by Éditions du Seuil

I *Le Phénomène humain*, Seuil, 1955
II *L'Apparition de l'Homme*, Seuil, 1956
III *La Vision du Passé*, Seuil, 1957
IV *Le Milieu divin*, Seuil, 1957
V *L'Avenir de l'Homme*, Seuil, 1959
VI *L'Énergie humaine*, Seuil, 1962
VII *L'Activation de l'Énergie*, Seuil, 1963
VIII *La Place de l'Homme dans la Nature (Le Groupe zoologique humain)*, Seuil, 1965
IX *Science et Christ*, Seuil, 1965
X *Comment je crois*, Seuil, 1969
XI *Les Directions de l'Avenir*, Seuil, 1973
XII *Écrits du temps de la guerre*, Seuil, 1975
XIII *Le Cœur de la matière*, Seuil, 1976

2 Collected Works in English translation

Dates shown below are the dates of publication by English-language publishers

I *The Human Phenomenon*, Sussex Academic Press (UK and US), 1999
II *The Appearance of Man*, Collins (UK), Harper and Row (US), 1965
III *The Vision of the Past*, Collins (UK), Harper and Row (US), 1966
IV *The Divine Milieu*, Collins (UK), Harper and Brothers (US), 1960
V *The Future of Man*, Collins (UK), Harper and Row (US), 1964
VI *Human Energy*, Collins (UK), Harcourt Brace Jovanovich (US), 1969
VII *Activation of Energy*, Collins (UK), Harcourt Brace Jovanovich (US), 1970
VIII *Man's Place in Nature*, Collins (UK), Harper and Row (US), 1966
IX *Science and Christ*, Collins (UK), Harper and Row (US), 1968

X *Christianity and Evolution*, Collins (UK), Harcourt Brace Jovanovich (US), 1971
XI *Toward the Future*, Collins (UK), Harcourt Brace Jovanovich (US), 1975
XII *Writings in Time of War*, Collins (UK), Harper and Row (US), 1968
XIII *The Heart of Matter*, Collins (UK), Harcourt Brace Jovanovich (US), 1978

3 Essays

Dates shown below are dates of composition or correction by Teilhard himself. Roman numerals indicate the volume number in the Seuil edition. Pages numbers are followed by "E" for the English text and "F" for the French. French titles were chosen by Teilhard himself or by his French editors. More than forty essay titles are incorrectly translated in the Collins/Harper/Harcourt English edition. Corrected English titles are given in square brackets.

Action and Activation, 9 August 1945, IX, 174 E; 219 F
Activation of Human Energy (The), 6 December 1953, VII, 385 E; 407 F
Africa and Human Origins, September 1954, II, 196 E; 275 F
Analysis of Life (The), 10 June 1945, VII, 129 E; 135 F
At the Wedding of Odette Bacot and Jean Teillard d'Éyry, 14 June 1928, XIII, 135 E; 157 F
At the Wedding of Éliane Basse and Hervé de la Goublaye de Ménorval, 15 March 1935, XIII, 139 E; 165 F
At the Wedding of Christine Dresch and Claude-Marie Haardt, 21 December 1948, XIII, 150 E; 187 F
Atomism of Spirit (The), 13 September 1941, VII, 21 E; 27 F
Australopithecines and the Missing Link in Evolution, June 1950, II, 126 E; 175 F
Australopithecines, Pithecanthropians and the Phyletic Structure of the Hominians, 21 January 1952, II, 177 E; 243 F
Awaited Word (The), 31 October 1940, XI, 92 E; 99 F

Basis and Foundations of the Idea of Evolution (The), 14 May 1926, III, 116 E; 163 F
Basis of My Attitude (The), 7 October 1948, XIII, 147 E; 179 F

Can Biology, Taken to its Extreme Limit, Enable Us to Emerge into the Transcendent?, May 1951, IX, 212 E; 277 F
Can Moral Science dispense with a Metaphysical Foundation? [Can Morality dispense with Metaphysical Foundations?], 23 April 1945, XI, 130 E; 141 F
Catholicism and Science, August 1946, IX, 187 E; 235 F
Centrology, 13 December 1944, VII, 97 E; 103 F
Christ in Matter, 14 October 1916, XIII, 61 E; 75 F
Christ the Evolver, 8 October 1942, X, 138 E; 161 F
Christian Phenomenon (The), 10 May 1950, X, 199 E; 231 F
Christianity and Evolution, 11 November 1945, X, 173 E; 201 F

Christianity in the World, May 1933, IX, 98 E; 129 F
Christic (The), March 1955, XIII, 80 E; 80 F
Christology and Evolution, 25 December 1933, X, 76 E; 93 F
Complementary Remarks on the Nature of Omega Point [Complementary Remarks on the Nature of Point Omega], 25 March 1954, II, 271 E; 371 F
Contingence of the Universe and Man's Zest for Survival (The) [The Contingency of the Universe and the Human Zest for Survival], 1 May 1953, X, 221 E; 263 F
Convergence of the Universe (The), 23 July 1951, VII, 281 E; 293 F
Cosmic Life, 24 April 1916, XII, 13 E; 17 F
Creative Union, November 1917, XII, 151 E; 193 F

Death-barrier and Co-reflection (The), 1 January 1955, VII, 395 E; 417 F
Defense of Orthogenesis in the Matter of Patterns of Speciation (A), January 1955, III, 268 E; 268 F
Degrees of Scientific Certainty in the Idea of Evolution (The), 15–20 November 1946, IX, 192 E; 243 F
Directions and Conditions of the Future (The), 30 June 1948, V, 227 E; 291 F
Discovery of Sinanthropus (The), 5 July 1937, II, 84 E; 119 F
Discovery of the Past (The), 15 September 1935, III, 183 E; 257 F
Divine Milieu (The), November 1926–March 1927, 1932, IV, 11 E; 17 F
Does Mankind Move Biologically Upon Itself? [Does Humanity Move Biologically Upon Itself?], 4 May 1949, V, 244 E; 317 F

Ecumenism, 15 December 1946, IX, 197 E; 251 F
End of the Species (The), 9 December 1952, V, 298 E; 387 F
Energy of Evolution (The), 24 May 1953, VII, 359 E; 379 F
Essence of the Democratic Idea (The) [The Essence of the Idea of Democracy], 2 February 1949, V, 238 E; 307 F
Eternal Feminine (The), 19–25 March 1918, XII, 191 E; 279 F
Evolution of Chastity (The), February 1934, XI, 60 E; 65 F
Evolution of Responsibility (The), May-June 1950, VII, 205 E; 211 F
Evolution of the Idea of Evolution (The), June-July 1950, III, 245 E; 345 F

Face of the Earth (The), 5–20 December 1921, III, 26 E; 41 F
Faith in Man [Faith in Humanity], February 1947, V, 185 E; 233 F
Faith in Peace, January 1947, V, 149 E; 189 F
Fall, Redemption and Geocentrism, 20 July 1920, X, 36 E; 47 F
Forma Christi, 13 December 1918, XII, 249 E; 363 F
Formation of the Noosphere (The), January 1947, V, 155 E; 199 F
Fossil Men [Fossil Humans], 20 March 1921, II, 25 E; 39 F
From the Pre-human to the Ultra-human, 27 April 1950, V, 289 E; 375 F
From Cosmos to Cosmogenesis, 15 March 1951, VII, 251 E; 259 F
Function of Art as an Expression of Human Energy (The) [How to Understand and Use Art as an Expression of Human Energy], 13 March 1939, XI, 88 E; 93 F

God of Evolution (The), 25 October 1953, X, 237 E; 283 F
Grand Option (The), 3 March 1939, V, 37 E; 55 F
Great Monad (The), 15 January 1918, XIII, 182 E; XII, 261 F

Heart of Matter (The), 15 August–30 October 1950, XIII, 15 E; 19 F
Heart of the Problem (The), 8 September 1949, V, 260 E; 337 F
Hominization, 6 May 1925, III, 51 E; 51 F
Hominization and Speciation, November-December 1952, III, 256 E; 363 F
How I Believe, 28 October 1934, X, 96 E; 115 F
How May We Conceive and Hope that Human Unanimization will be realized on Earth?, 18 January 1950, V, 281 E; 365 F
How the Transformist Question Presents Itself Today, 5–20 June 1921, III, 7 E; 15 F
Human Energy, 6 August-8 September 1937, VI, 113 E; 141 F
Human Phenomenon (The), June 1938–June 1940, I E; 21 F
Human Rebound of Evolution (The), 23 September 1947, V, 196 E; 251 F

Introduction to the Christian Life, 29 June 1944, X, 151 E; 177 F

Last Page of Pierre Teilhard de Chardin's Journal (The) [The Last Page of Pierre Teilhard de Chardin's Diary], 7 April 1955, XIII, 103 E; 119 F
Law of Irreversibility in Evolution (The), 21 March 1923, III, 49 E; 71 F
Life and the Planets, 10 March 1945, V, 97 E; 127 F

Major Problem for Anthropology (A), 30 December 1951, VII, 311 E; 325 F
Man's Place in Nature [The Human Place in Nature], 1932, III, 175 E; 245 F
Man's Place in Nature [The Human Zoological Group], 4 August 1949, VIII, 14 E; 19 F
Man's Place in the Universe [The Human Place in the Universe], 15 November 1942, III, 216 E; 303 F
Mass on the World (The), 1923, XIII, 119 E; 139 F
Mastery of the World and the Kingdom of God, 20 September 1916, XII, 73 E; 83 F
Modern Unbelief, 25 October 1933, IX, 113 E; 147 F
Moment of Choice (The), Christmas 1939, VII, 11 E; 17 F
Monogenism and Monophyletism, 1950, X, 209 E; 245 F
Movements of Life (The), April 1928, III, 143 E; 199 F
My Fundamental Vision [How I See], 12 August 1948, XI, 163 E; 177 F
My Fundamental Vision [How I See], Appendix, 26 August 1948, XI, 206 E; 221 F
My Intellectual Position, April 1948, XIII, 143 E; 171 F
My Litany, October(?) 1953, X, 244 E; 293 F
My Universe, 14 April 1918, XIII, 196 E; XII, 293 F
My Universe, 25 March 1924, IX, 37 E; 63 F
Mystical Milieu (The), 13 August 1917, XII, 115 E; 153 F
Mysticism of Science (The), 20 March 1939, VI, 163 E; 201 F

Names of Matter (The), Easter 1919, XIII, 225 E; XII, 447 F
Natural History of the World (The), January 1925, III, 103 E; 143 F
Natural Units of Humanity (The) [The Natural Human Units], 5 July 1939, III, 192 E; 271 F
Necessarily Discontinuous Appearance of Every Evolutionary Series (The), 17 March 1926, III, 114 E; 159 F
New Spirit (The), 13 February 1942, V, 82 E; 107 F
Nostalgia for the Front, September 1917, XIII, 167 E; XII, 225 F
Note on Progress (A), 17 September 1920, V, 11 E; 21 F
Note on Some Possible Historical Representations of Original Sin, 15(?) April 1922, X, 45 E; 59 F
Notes on South African Prehistory, November(?) 1951, II, 172 E; 235 F
Note on the Biological Structure of Mankind [Note on the Biological Structure of Humanity], 3 August 1948, IX, 206 E; 265 F
Note on the Concept of Christian Perfection (A) [A Note on the Notion of Christian Perfection], 1942, XI, 101 E; 111 F
Note on the Essence of Transformism, 1920, XIII, 107 E; 123 F
Note on the Modes of Divine Action in the Universe, January 1920, X, 25 E; 33 F
Note on the Physical Union between the Humanity of Christ and the Faithful in the Course of their Sanctification, January(?) 1920, X, 15 E; 19 F
Note on the Present Reality and Evolutionary Significance of a Human Orthogenesis, 5 May 1951, III, 248 E; 351 F
Note on the Presentation of the Gospel in a New Age [Note on the Evangelization of the New Times], 6–10 January 1919, XIII, 209 E; XII, 395 F
Note on the Universal Christ, January 1920, IX, 14 E; 37 F
Note on the Universal Element of the World, 22 December 1918, XII, 271 E; 387 F

Observations on the Australopithecines, March 1952, II, 180 E; 249 F
Observations on the Teaching of Prehistory, 23 September 1948, XIII, 145 E; 175 F
On Looking at a Cyclotron, April 1953, VII, 347 E; 365 F
On My Attitude to the Official Church, 5 January 1921, XIII, 115 E; 133 F
On the Nature of the Phenomenon of Human Society [On the Nature of the Human Social Phenomenon], 23 April 1948, VII, 165 E; 171 F
On the Notion of Creative Transformation, 1919 (or early 1920), X, 21 E; 27 F
On the Probability of an Early Bifurcation of the Human Phylum, 23 November 1953, II, 185 E; 257 F
On the Probable Coming of an "Ultra-human" [On the Probable Existence Ahead of an "Ultra-human"], 6 January 1950, V, 270 E; 351 F
Operative Faith, 28 September 1918, XII, 225 E; 335 F
Outline of a Dialectic of Spirit, 25 November 1946, VII, 141 E; 147 F

Paleontology and the Appearance of Man [Paleontology and the Appearance of Humanity], March–April 1923, II, 33 E; 51 F

Pantheism and Christianity, 17 January 1923, X, 56 E; 71 F
Phenomenon of Counter-Evolution in Human Biology (A), 26 January 1949, VII, 181 E; 187 F
Phenomenon of Man (The) [The Human Phenomenon], September 1928, IX, 86 E; 115 F
Phenomenon of Man (The) [The Human Phenomenon], November 1930, III, 161 E; 225 F
Phenomenon of Man (The) [The Human Phenomenon], June 1954, XIII, 155 E; 197 F
Phenomenon of Man (The) [The Human Phenomenon]: An Essential Observation, 17 October 1948, XIII, 149 E; 183 F
Phenomenon of Spirituality (The) [The Spiritual Phenomenon], March 1937, VI, 93 E; 115 F
Planetization of Mankind (The) [Human Planetization], 25 December 1945, V, 124 E; 157 F
Phyletic Structure of the Human Group (The), February 1951, II, 132 E; 185 F
Place of Technology in a General Biology of Mankind (The) [The Place of Technology in a General Biology of Humanity], 16 January 1947, VII, 153 E; 159 F
Pleistocene Fauna and the Age of Man in North America (The) [The Pleistocene Fauna and the Age of Humanity in North America], 1935, II, 79 E; 111 F
Plurality of Inhabited Worlds (The), 5 June 1953, X, 229 E; 273 F
Prehistoric Excavations of Peking (The), 20 March 1934, II, 68 E; 97 F
Progress of Prehistory, 5 January 1913, II, 11 E; 21 F
Promised Land (The), February 1919, XII, 277 E; 415 F
Priest (The), 8 July 1918, XII, 203 E; 309 F
Psychological Conditions of the Unification of Man (The) [The Psychological Conditions of Human Unification], 6 January 1949, VII, 169 E; 175 F

Qualifications, Career, Field-work and Writings of Pierre Teilhard de Chardin [Qualifications and Writings of Pierre Teilhard de Chardin], September 1948, XIII, 157 E; 201 F
Question of Fossil Man (The) [The Question of Fossil Humans], 15 September 1943, II, 93 E; 133 F

Reflection of Energy (The), 27 April 1952, VII, 319 E; 333 F
Reflections on Happiness, 28 December 1943, XI, 107 E; 119 F
Reflections on the Compression of Mankind [Reflections on Human Compression], 18 January 1953, VII, 339 E; 355 F
Reflections on Original Sin, 15 November 1947, X, 187 E; 217 F
Reflections on Progress, 22 February–30 March 1941, V, 61 E; 83 F
Reflections on the Scientific Probability and the Religious Consequences of an Ultra-human, 25 March 1951, VII, 269 E; 279 F
Reflections on Two Converse Forms of Spirit [Reflections on Two Inverse Forms of Spirit], 25 July 1950, VII, 215 E; 223 F
Religious Value of Research (The) [On the Religious Value of Research], 20 August 1947, IX, 199 E; 255 F

Research, Work and Worship, March 1955, IX, 214 E; 281 F
Rise of the Other (The), 20 January 1942, VII, 59 E; 65 F
Road of the West (The), 8 September 1932, XI, 40 E; 45 F

Salvation of Mankind (The) [The Salvation of Humanity], 11 November 1936, IX, 128 E; 167 F
Science and Christ, 27 February 1921, IX, 21 E; 45 F
Scientific Career of Pierre Teilhard de Chardin, July–August 1950, XIII, 152 E; 191 F
Search for the Discovery of Human Origins south of the Sahara, June 1954, II, 188 E; 263 F
Sense of Man (The) [Human Sense], February–March 1929, XI, 13 E; 19 F
Sense of the Cross (The), 14 September 1952, X, 212 E; 251 F
Sense of the Species (The), 31 May 1949, VII, 197 E; 203 F
Significance and Positive Value of Suffering (The), 1 April 1933, VI, 48 E; 59 F
Sinanthropus Pekinensis, April 1930, II, 58 E; 83 F
Singularities of the Human Species (The), 25 March 1954, II, 208 E; 293 F
Singularity of the Christian Phenomenon ["Unique Nature of the Christian Phenomenon"], 25 March 1954, II, 271 E; 371 F
Sketch of a Personalistic Universe [Outline of a Personalistic Universe], 4 May 1936, VI, 53 E; 67 F
Social Heredity and Progress, 1938, V, 25 E; 39 F
Some Notes on the Mystical Sense [Some Clarifying Remarks on the Mystical Sense], Winter 1951, XI, 209 E; 225 F
Some General Views on the Essence of Christianity, May 1939, X, 133 E; 153 F
Some Reflections on the Conversion of the World, 9 October 1936, IX, 118 E; 155 F
Some Reflections on the Rights of Man [Some Reflections on Human Rights], 22 March 1947, V, 193 E; 245 F
Some Reflections on the Spiritual Repercussions of the Atom Bomb, September 1946, V, 140 E; 177 F
Soul of the World (The), Epiphany, 6 January 1918, XII, 177 E; 243 F
Spirit of the Earth (The), 9 March 1931, VI, 19 E; 23 F
Spiritual Contribution of the Far East (The), 10 February 1947, XI, 134 E; 147 F
Spiritual Energy of Suffering (The), 8 January 1950, VII, 245 E; 253 F
Spiritual Power of Matter (The), 8 August 1919, XIII, 67 E; XII, 465 F; XIII, 81 F
Struggle against the Multitude (The), 26 February–22 March 1917, XII, 93 E; 129 F
Stuff of the Universe (The), 14 July 1953, VII, 373 E; 395 F
Summary of My Phenomenological View of the World, 14 January 1954, XI, 212 E; 231 F
Super-Humanity, Super-Christ, Super-Charity, August 1943, IX, 151 E; 193 F

Transformation and Continuation in Man of the Mechanism of Evolution (The) [The Transformation and Continuation in Humanity of the Mechanism of

Evolution], 19 November 1951, VII, 297 E; 311 F
Transformist Paradox (The), January 1925, III, 80 E; 113 F
Turmoil or Genesis? [Agitation or Genesis?], 20 December 1947, V, 214 E; 273 F
Two Principles and a Corollary [Three Things I See], February 1948, XI, 148 E; 161 F

Universal Element (The), 21 February 1919, XII, 289 E; 429 F
Universalization and Union, 20 March 1942, VII, 77 E; 83 F

Vision of the Past (The), 17–22 October 1949, III, 237 E; 333 F

What exactly is the Human Body?, August(?) 1919, IX, 11 E; 31 F
What is Life?, 2 March 1950, IX, 210 E; 273 F
What Should We Think of Transformism?, January 1930, III, 151 E; 211 F

Zest for Living (The), November 1950, VII, 229 E; 237 F
Zoological Evolution and Invention, April 1947, III, 234 E; 327 F

4 Letters in French

Accomplir l'homme, lettres inédites, 1926–52, Grasset, 1968
Blondel et Teilhard de Chardin, 1919, Beauchesne, 1965
Genèse d'une pensée, 1914–19, Grasset, 1961
Lettres à Jeanne Mortier, 1939–55, Seuil, 1984
Lettres à Léontine Zanta, 1923–39, Desclée de Brouwer, 1965
Lettres d'Égypte, 1905–08, Aubier, 1963
Lettres d'Hastings et de Paris, 1908–14, Aubier, 1965
Lettres familières de Pierre Teilhard de Chardin, 1948–55, Centurion, 1976
Lettres inédites à Christophe Gaudefroy et Henri Breuil, 1923–55, Rocher, 1988
Lettres intimes de Teilhard de Chardin à Auguste Valensin, Bruno de Solages, Henri de Lubac, André Ravier, 1919–55, Aubier, 1974
Lettres de voyage, 1923–55, Grasset, 1956, 1967, Maspero, 1982

5 Letters in English or in English translation

Letters from a Traveler, 1923–55, Collins (UK), Harper and Row (US), 1962
Letters from Egypt, 1905–08, Herder and Herder, 1965
Letters from Hastings, 1908–12, Herder and Herder, 1968
Letters from My Friend Teilhard de Chardin, 1948–55, Paulist Press, 1980
Letters from Paris, 1912–14, Herder and Herder, 1967
Letters of Teilhard de Chardin and Lucile Swan, 1932–55, Georgetown University Press, 1993
Letters to Léontine Zanta, 1923–39, Collins (UK), Harper and Row (US), 1969
Letters to Two Friends, 1926–52, New American Library (US), 1968, Fontana (UK), 1972

Pierre Teilhard de Chardin-Maurice Blondel, Correspondence, 1919, Herder and Herder, 1967
The Making of a Mind, 1914–19, Collins (UK), Harper and Row (US), 1965

6 Other texts in French

Je m'explique, Seuil, 1966
Hymne de l'univers, Seuil, 1961
Pierre Teilhard de Chardin, Journal, 1915–19, Fayard, 1975
Réflexions et Prières dans l'Espace-Temps, Édouard et Suzanne Bret, edd., Seuil, 1972
Sur l'Amour, Seuil, 1967
Sur le Bonheur, Seuil, 1966
Sur la Souffrance, Seuil, 1974

7 Other texts in English

Hymn of the Universe, Collins, (UK), Harper and Row (US), 1965
Let Me Explain, Collins (UK), Harper and Row (US), 1970
On Happiness, Collins (UK), Harper and Row (US), 1973
On Love, Collins (UK), Harper and Row (US), 1972
On Suffering, Collins (UK), Harper and Row (US), 1975

SECONDARY SOURCES

8 Teilhard's language

BARTHÉLEMY-MADAULE, Madeleine, "Tableaux des catégories teilhardiennes", in *Bergson et Teilhard de Chardin*, Seuil, 1963
CUÉNOT, Claude, *Nouveau Lexique Teilhard de Chardin*, Seuil, 1968
CUÉNOT, Claude, "Vocabulaire", in *Teilhard de Chardin*, Édition Écrivains de toujours, Seuil, 1962
CUYPERS, Hubert, *Vocabulaire Teilhard*, Presses Universitaires, 1963
HAAS, Adolf, SJ, *Teilhard de Chardin-Lexikon, Grundbegriffe, Erläuterungen, Texte*, 2 Vols., Herderbücherei, 1971
KLAUDER, Francis, SDB, "Glossary of Interrelated Terms", in *Aspects of the Thought of Teilhard de Chardin*, Christopher, 1971
L'ARCHEVÊQUE, Paul, *Teilhard de Chardin, Nouvel index analytique*, Presses de l'Université Laval, 1972
RIDEAU, Émile, SJ, "Le vocabulaire et le langage", in *La Pensée du Père Teilhard de Chardin*, Seuil, 1965; "Vocabulary and Language", in *Teilhard de Chardin, A Guide to His Thought*, Collins, 1967
RUSSO, François, SJ, "Les Langages du Père Teilhard de Chardin", in *Teilhard de Chardin, son apport, son actualité*, Colloque du Centre Sèvres, Centurion, 1981

9 Select biographies and other studies

Books in bold are essential introductory reading for English-speaking users: many were still in print at the time of going to press.

BARBOUR, George, *In the Field with Teilhard de Chardin*, Herder and Herder, 1965

BARJON, Louis, SJ, *Le Combat de Pierre Teilhard de Chardin*, Presses de l'Université Laval, 1971

BARJON, Louis, SJ, et LEROY, Pierre, SJ, *La Carrière scientifique du Père Teilhard de Chardin*, Rocher, n.d.

BARTHÉLEMY-MADAULE, Madeleine, *Bergson et Teilhard de Chardin*, Seuil, 1963

BARTHÉLEMY-MADAULE, Madeleine, *La Personne et le drame humain chez Teilhard de Chardin*, Seuil, 1967

BAUDRY, Gérard-Henry, *Ce que croyait Teilhard*, Mame, 1971

BAUDRY, Gérard-Henry, *Qui était Teilhard de Chardin*, Baudry, 1972

BAUDRY, Gérard-Henry, *Dictionnaire des correspondants de Teilhard de Chardin*, 1892–1955, Baudry, 1974

BAUDRY, Gérard-Henry, *Le Christ universel, Espoir pour le Monde*, Baudry, 1979

BERGERON, Marie-Ina, SMM, *La Chine et Teilhard*, Delarge, 1976

BRAYBROOKE, Neville, ed., *Teilhard de Chardin, Pilgrim of the Future*, Darton, Longman and Todd, 1965

CARLES, Jules, et DUPLEIX, André, *Pierre Teilhard de Chardin*, Centurion, 1991

CHAUCHARD, Paul, *La Foi du savant chrétien*, Aubier, 1965

CORBISHLEY, Thomas, SJ, *The Spiritual Exercises of St Ignatius*, Anthony Clarke, 1963

CORBISHLEY, Thomas, SJ, *The Spirituality of Teilhard de Chardin*, Fontana, 1971

COUSINS, Ewert, ed., *Hope and the Future of Man*, Teilhard Study Library, Teilhard Centre/Garnstone, 1973

COUSINS, Ewert, *Christ of the 21st Century*, Element, 1992

CRESPY, Georges, *La pensée théologique de Teilhard de Chardin*, Éditions Universitaires, 1961

CUÉNOT, Claude, *Pierre Teilhard de Chardin*, Plon, 1958, Rocher, 1986; *Teilhard de Chardin*, Burns and Oates, 1965

CUÉNOT, Claude, *Teilhard de Chardin*, Édition Écrivains de toujours, Seuil, 1962

CUÉNOT, Claude, *Science and Faith in Teilhard de Chardin*, Teilhard Study Library, Garnstone, 1967

DALEUX, André, *Teilhard de Chardin, Science et foi réconciliées?*, Éditions Gabriandre, 1994

DELFGAAUW, Bernard, *Teilhard de Chardin*, Wereldfenster, 1961; Evolution, *The Theory of Teilhard de Chardin*, Fontana, 1969

DE TERRA, Helmut, *Memories of Teilhard de Chardin*, Collins, 1964

DOBZHANSKY, Theodosius, *The Biology of Ultimate Concern*, New American Library, 1967, Fontana, 1971

DODSON, Edward, *The Phenomenon of Man Revisited*, Columbia University Press, 1984
D'OUINCE, René, SJ, *Un prophète en procès*, 2 vols., Aubier, 1970
DUPLEIX, André, *Prier 15 jours avec Pierre Teilhard de Chardin*, Nouvelle Cité, 1994
Faculté de Théologie de Lyon-Fourvière, *L'Homme devant Dieu, Mélanges offerts au Père Henri de Lubac*, Aubier, 1964
FARICY, Robert, SJ, *Building God's World*, Dimension, 1976
FARICY, Robert, SJ, *Christian Faith and My Everyday Life, The Spiritual Doctrine of Teilhard de Chardin*, St. Paul, 1981
FARICY, Robert, SJ, *All Things in Christ, Teilhard de Chardin's Spirituality*, Fount, 1981
FARICY, Robert, SJ, and ROONEY, Lucy, SND, *Knowing Jesus in the World, Praying with Teilhard de Chardin*, St Pauls, 1999
GALLAGHER, Blanche Marie, BVM, *Meditations with Teilhard de Chardin*, Bear, 1988
GALLYON, Margaret, *The Visions, Revelations and Teachings of Angela of Foligno*, Sussex Academic Press, 2000
GIRET, Raoul, *Teilhard aujourd'hui*, Aubin, 1996
GRAU, Joseph, *Morality and the Human Future in the Thought of Teilhard de Chardin*, Fairleigh Dickinson, 1976
GRAY, Donald, *The One and the Many*, Burns and Oates, 1969
GRENET, Paul, *Teilhard de Chardin*, Souvenir, 1965
HAAS, Giulio, *Die Weltsicht von Teilhard und Jung, Gegensätze, die sich vereinen*, Walter, 1991
HANSON, Anthony, ed., *Teilhard Reassessed*, Darton, Longman and Todd, 1970
KING, Thomas, SJ, *Teilhard's Mysticism of Knowing*, Seabury, 1981
KING, Thomas, SJ, *Teilhard de Chardin, The Way of the Christian Mystics*, Michael Glazier, 1988
KING, Ursula, *Towards a New Mysticism*, Collins, 1980 (UK), Seabury (US), 1980
KING, Ursula, *The Spirit of One Earth, Reflections on Teilhard de Chardin and Global Spirituality*, Paragon, 1989
KING, Ursula, *Spirit of Fire, The Life and Vision of Teilhard de Chardin*, Orbis, 1996
KING, Ursula, *Christ in All Things, Exploring Spirituality with Teilhard de Chardin*, SCM, 1997
KING, Ursula, *Pierre Teilhard de Chardin, Selected Writings*, Orbis, 1999
KLAUDER, Francis, SDB, *Aspects of the Thought of Teilhard de Chardin*, Christopher, 1971
KROPF, Richard, *Teilhard, Scripture and Revelation*, Fairleigh Dickinson, 1980
LABERGE, Jacques, SJ, *Pierre Teilhard de Chardin et Ignace de Loyola, Notes de retraite*, 1919–55, Desclée de Brouwer, 1973
LEROY, Pierre, SJ, *Pierre Teilhard de Chardin tel que je l'ai connu*, Plon, 1958
LEROY, Pierre, SJ, *Un Chemin non tracé, Jésuite au XXe siècle*, Desclée de Brouwer, 1992

LEROY, Pierre, SJ, MORIN, Hélène, et SOULIÉ, Solange, *Pèlerin de l'Avenir, Le Père Teilhard de Chardin à travers sa correspondance*, 1905–55, Centurion, 1989

LEWIS, John, ed., *Beyond Chance and Necessity*, Teilhard Study Library, Teilhard Centre/Garnstone, 1974

LISCHER, Richard, *Marx and Teilhard, Two Ways to a New Humanity*, Orbis, 1979

de LUBAC, Henri, SJ, *La Pensée religieuse du Père Teilhard de Chardin*, Montaigne, 1962; *The Religion of Teilhard de Chardin*, Collins, 1967

de LUBAC, Henri, SJ, *La Prière du Père Teilhard de Chardin*, Fayard, 1964; *The Faith of Teilhard de Chardin*, Burns Oates, 1965

de LUBAC, Henri, SJ, ed., *Blondel et Teilhard de Chardin, Correspondance*, Beauchesne, 1965; *Pierre Teilhard de Chardin-Maurice Blondel, Correspondence*, Herder and Herder, 1967

de LUBAC, Henri, SJ, *L'Éternel Féminin*, Aubier, 1968; *The Eternal Feminine*, Collins, 1971

de LUBAC, Henri, SJ, *Teilhard et notre temps*, Aubier, 1971; "Teilhard and the Problems of Today", in *The Eternal Feminine*, Collins, 1971

de LUBAC, Henri, SJ, *Teilhard Posthume, Réflexions et souvenirs*, Fayard, 1977

LUKAS, Mary, and LUKAS, Ellen, *Teilhard*, Collins, 1977

LYONS, James, SJ, *The Cosmic Christ in Origen and Teilhard de Chardin*, Oxford University Press, 1982

McCARTHY, Joseph, *Pierre Teilhard de Chardin, A Comprehensive Bibliography*, Garland, 1981

MALONEY, George, SJ, *The Cosmic Christ from Paul to Teilhard*, Sheed and Ward, 1968

MALONEY, George, SJ, *The Breath of the Mystic*, Dimension, 1974

MALONEY, George, SJ, *Mysticism and the New Age, Christic Consciousness in the New Creation*, Alba House, 1991

MATHIEU, Pierre-Louis, *La pensée politique et économique de Teilhard de Chardin*, Seuil, 1969

MOONEY, Christopher, SJ, *Teilhard de Chardin and the Mystery of Christ*, Collins, 1966

MOURGUE, Gérard, *Sri Aurobindo et Teilhard de Chardin*, Buchet/Chastel, 1993

O'MANIQUE, John, *Energy in Evolution*, Teilhard Study Library, Garnstone, 1969

ONIMUS, Jean, *Teilhard de Chardin et le mystère de la Terre*, Albin Michel, 1991

RAVEN, Charles, *Teilhard de Chardin, Scientist and Seer*, Collins, 1962

RAVIER, André, SJ, "Teilhard de Chardin et l'expérience mystique d'après ses notes intimes", in *Terre promise*, Cahiers Pierre Teilhard de Chardin, N° 8, Fondation et Association Teilhard de Chardin, Seuil, 1974

REYNOLDS, Barbara, intro., Dante, *The Divine Comedy, Paradise*, Penguin, 1962,

RIDEAU, Émile, SJ, *La Pensée du Père Teilhard de Chardin*, Seuil, 1965; Teilhard de Chardin, *A Guide to His Thought*, Collins, 1967

SIPRIOT, Pierre, ed., *Pierre Teilhard de Chardin, naissance et avenir de l'Homme*, Les Cahiers du Rocher, Rocher, 1987

de SOLAGES, Bruno, OCarm, "Les preuves teilhardiennes de Dieu", in *L'homme devant Dieu, Mélanges offerts au Père Henri de Lubac*, Vol. III, Aubier, 1964

de SOLAGES, Bruno, OCarm, *Teilhard de Chardin, Témoignage et étude sur le développement de sa pensée*, Privat, 1967

SPEAIGHT, Robert, *Teilhard de Chardin*, Collins, 1967

SZEKERES, Attila, ed., *Le Christ cosmique de Teilhard de Chardin*, Seuil, 1969

TOWERS, Bernard, intro., *Evolution, Marxism and Christianity*, Teilhard Study Library, Garnstone, 1967

TOWERS, Bernard, *Concerning Teilhard and Other Writings on Science and Religion*, Collins, 1969

WILDIERS, Norbertus Maximiliaan, OFMCap, *Teilhard de Chardin, Een inleiding in zijn denken*, Standaard, 1963; *Teilhard de Chardin, An Introduction*, Fontana, 1968

ZONNEVELD, Leo, ed., *The Desire to be Human*, International Teilhard Compendium, Mirananda, 1983

10 Colloquia and Symposia

Teilhard de Chardin et la pensée catholique, Colloque de Venise, Claude Cuénot, ed., Seuil, 1965

Teilhard de Chardin, son apport, son actualité, Colloque du Centre Sèvres, 1981, Centurion, 1982

Teilhard and the Unity of Knowledge, Georgetown University Centennial Symposium, 1981, Thomas King SJ and James Salmon SJ, edd., Paulist Press, 1983

Humanity's Quest for Unity, A United Nations Teilhard Symposium, Leo Zonneveld, ed., United Nations University for Peace, Mirananda, 1985

11 Selected publications of the American Teilhard Association

(a) *Teilhard Studies*

ANDERSON, Irvine, *History in a Teilhardian Context: The Thought of Teilhard de Chardin as a Guide to Social Science*, No. 17, Anima, Spring/Summer 1987

BALTAZAR, Eulalio, *Liberation Theology and Teilhard de Chardin*, No. 20, Anima, Fall/Winter 1988

BERRY, Thomas, CP, *The New Story: Comments on the Origin, Identification and Transmission of Values*, No. 1, Anima, Winter 1978

BERRY, Thomas, CP, *Management: The Managerial Ethos and the Future of Planet Earth*, No. 3, Anima, Spring 1980

BERRY, Thomas, CP, *Technology and the Healing of the Earth*, No. 14, Anima, Fall 1985

DODSON, Edward, *Teilhard and Mendel: Contrasts and Parallels*, No. 12, Anima, Fall 1984

DODSON, Edward, *The Teilhardian Synthesis, Lamarckism, and Orthogenesis*, No. 29, Anima, Summer 1993

DUPUY, Kenneth, *The Once and Future Earth: The Millennarian Programmatic in Bacon and Teilhard*, No. 39, American Teilhard Association, Spring 2000

FABEL, Arthur, *The New Book of Nature*, No. 8, Anima, Fall 1982

FABEL, Arthur, *Cosmic Genesis: Teilhard de Chardin and the Emerging Scientific Paradigm*, No. 5, Anima, Summer 1981

FABEL, Arthur, *Teilhard 2000, The Vision of a Cosmic Genesis at the Millennium*, No. 36, American Teilhard Association, Spring 1998

FALLA, William, *Syntheses in Science and Religion: A Comparison*, No. 35, American Teilhard Association, Summer 1997

GRAU, Joseph, *The Creative Union of Person and Community: A Geo-Humanist Ethic*, No. 22, Anima, Fall/Winter 1989

GRAY, Donald, *A New Creation Story: The Creative Spirituality of Teilhard de Chardin*, No. 2, Anima, Spring 1979

GRIM, John, *Reflections on Shamanism: The Tribal Healer and the Technological Trance*, Anima, Fall 1981

GRIM, John, *Apocalyptic Spirituality in the Old and New Worlds: The Revisioning of History and Matter*, No. 27, Anima, Autumn 1992

GRIM, John, and GRIM, Mary Evelyn, *Teilhard de Chardin: A Short Biography*, No. 11, Anima, Spring 1984

HAUGHT, John, *Chaos, Complexity and Theology*, No. 30, Anima, Summer 1994

HOFSTETTER, Adrian, OP, *The New Biology: Barbara McClintock and an Emerging Holistic Science*, No. 26, Anima, Spring 1992

KING, Thomas, SJ, *Teilhard's Unity of Knowledge*, No. 9, Anima, Summer 1983

KING, Thomas, SJ, *Teilhard, Evil and Providence*, No. 21, Anima, Spring/Summer 1989

KING, Ursula, *The Letters of Teilhard de Chardin and Lucile Swan*, No. 32, American Teilhard Association, Fall 1995

KRAFT, Wayne, *Love as Energy*, No. 19, Anima, Spring/Summer 1988

McCULLOCH, Winifred, *Teilhard de Chardin and the Piltdown Hoax*, No. 33, American Teilhard Association, Spring 1996

MAALOUF, Jean, *The Divine Milieu*, No. 38, American Teilhard Association, Autumn 1999

MARIE-DALY, Bernice, *Ecofeminism: Sacred Matter/Sacred Mother*, No. 25, Anima, Autumn 1991

MOONEY, Christopher, SJ, *Cybernation, Responsibility and Providential Design*, No. 24, Anima, Summer 1991

NICHOLS, Marilyn, SSJ, *The Journey Symbol*, No. 18, Anima, Fall/Winter 1987

NORRIS, Russel, *Creation, Cosmology, and the Cosmic Christ*, No. 31, American Teilhard Association, Spring 1995

O'HARA, Dennis Patrick, and ST. JOHN, Donald P., *Merton and Ecology*, No. 37, Winter/Spring 1999

REES, William, *Sustainable Development and the Biosphere*, No. 23, Anima, Spring/Summer 1990

RYAN, John, *Psychotherapy, Religion and the Teilhardian Vision,* No. 34, American Teilhard Association, Winter 1997
SALMON, James, SJ, *Teilhard and Prigogine*, No. 16, Anima, Fall/Winter 1986
STIKKER, Allerd, *Teilhard, Taoism, and Western Thought*, No. 15, Anima, Spring/Summer 1986
SWIMME, Brian, *The New Natural Selection*, No. 10, Anima, Fall 1983
TUCKER, Mary Evelyn, *The Ecological Spirituality of Teilhard*, No. 13, Anima, Spring 1985
TUCKER, Mary Evelyn, *Education and Ecology*, No. 28, Anima, Spring 1993
WOLSKY, Alexander, *Teilhard de Chardin's Biological Ideas*, No. 4, Anima, Spring 1981

(b) Other publications

BERRY, Thomas, CP, *Contemporary Spirituality*, Riverdale Studies No. 1, Riverdale, 1975
GRAY, Donald, *The Phenomenon of Man: Its Inner Dialectic*, Riverdale Studies No. 2, Riverdale, 1976

Acknowledgements

The author and publishers gratefully acknowledge the financial support of the French Minister for Culture – National Book Centre

The author and publishers gratefully acknowledge the following for permission to use copyright material:

David Higham Associates for excerpts from Dante, *The Divine Comedy*, translated by Barbara Reynolds

Desclée de Brouwer for excerpts from *Pierre Teilhard de Chardin et Ignace de Loyola* by Jacques Laberge

Éditions Bernard Grasset for excerpts from *Écrits du temps de la guerre* and *Genèse d'une pensée* by Pierre Teilhard de Chardin

Éditions Édouard Privat for excerpts from *Teilhard de Chardin* by Bruno de Solages

Éditions du Rocher for excerpts from *Pierre Teilhard de Chardin* by Claude Cuénot

Éditions du Seuil for excerpts from *L'apparition de l'homme* by Pierre Teilhard de Chardin © Éditions du Seuil, 1956; *La vision du passé* by Pierre Teilhard de Chardin © Éditions du Seuil, 1957; *L'avenir de l'homme* by Pierre Teilhard de Chardin © Éditions du Seuil, 1959; *L'énergie humaine* by Pierre Teilhard de Chardin © Éditions du Seuil, 1962; *Science et Christ* by Pierre Teilhard de Chardin © Éditions du Seuil, 1965

Harcourt Inc. for excerpts from *Activation of Energy* by Pierre Teilhard de Chardin, copyright © 1963 by Éditions du Seuil, English translation by René Hague copyright © 1970 by William Collins Sons & Co. Ltd., London, reprinted by permission of Harcourt Inc.

for excerpts from *Toward the Future* by Pierre Teilhard de Chardin, copyright © 1973 by

Éditions du Seuil, English translation by René Hague copyright © 1975 by William Collins Sons & Co. Ltd. and Harcourt Inc., reprinted by permission of Harcourt Inc.
for excerpts from *The Heart of Matter* by Pierre Teilhard de Chardin, copyright © 1976 by Éditions du Seuil, English translation copyright © 1978 by William Collins Sons & Co. Ltd. and Harcourt Inc., reprinted by permission of Harcourt Inc.

HarperCollins Publishers Inc. for excerpts from *The Divine Milieu* by Pierre Teilhard de Chardin copyright © 1957 by Éditions du Seuil, Paris. English translation copyright © by Wm. Collins Sons & Co., London, and Harper & Row, Publishers, Inc., New York. Renewed © 1988 by Harper & Row Publishers, Inc.

Librairie Arthème Fayard for excerpts from *Journal: 26 août 1915–14 janvier 1919* by Pierre Teilhard de Chardin © Librairie Arthème Fayard, 1975

Paulist Press for excerpts *Letters from My Friend Teilhard de Chardin* by Pierre Leroy

Sussex Academic Press for excerpts from *The Human Phenomenon* by Pierre Teilhard de Chardin. A new edition and translation by Sarah Appleton-Weber, copyright © 1999.

The publishers apologize for any errors or omissons in the above list and would be grateful to be notified of any corrections that should be incorporated in the next edition or reprint of this book.